T0205528

Progress in Mathematical Physics

Volume 76

More information about this series at http://www.springer.com/series/4813

Bertrand Duplantier • Vincent Rivasseau

Editors

The Universe

Poincaré Seminar 2015

 Birkhäuser

Editors
Bertrand Duplantier 🔟
CEA Saclay
Institut de Physique Théorique
Gif-sur-Yvette Cedex, France

Vincent Rivasseau
Laboratoire de Physique Théorique
Université Paris-Sud
Orsay, France

ISSN 1544-9998   ISSN 2197-1846   (electronic)
Progress in Mathematical Physics
ISBN 978-3-030-67394-9   ISBN 978-3-030-67392-5   (eBook)
https://doi.org/10.1007/978-3-030-67394-9

Mathematics Subject Classification (2020): 83-02, 83C57, 83F05, 85-02, 85A20, 85A40

This book is published under the imprint Birkhäuser, www.birkhauser-science.com by the registered company Springer Nature Switzerland AG
The registered company address is: Gewerbestrasse 11, 6330 Cham, Switzerland

Contents

Arnaud Cassan

Foreword

This book is the seventeenth in a series of Proceedings for the *Séminaire Poincaré*, which is directed towards a broad audience of physicists, mathematicians, and philosophers of science.

The goal of the Poincaré Seminar is to provide up-to-date information about general topics of great interest in physics. Both the theoretical and experimental aspects of the topic are covered, generally with some historical background. Inspired by the *Nicolas Bourbaki Seminar* in mathematics, hence nicknamed *"Bourbaphy"*, the Poincaré Seminar is held once or twice a year at the Institut Henri Poincaré in Paris, with written contributions prepared in advance. Particular care is devoted to the pedagogical nature of the presentations, so that they may be accessible to a large audience of scientists.

This new volume of the Poincaré Seminar Series, **The Universe**, corresponds to the twentieth such seminar, held on November 21, 2015, at Institut Henri Poincaré in Paris. Its aim is to provide a description of some of the main active areas in astrophysics from the largest scales probed by the *Planck* satellite to massive black holes which lie at the heart of galaxies and up to the much awaited but stunning discovery of thousands of exoplanets.

The first article, entitled *The Big-Bang Theory: Construction, Evolution and Status*, by the distinguished theoretical astrophysicist, JEAN-PHILIPPE UZAN, formerly Deputy Director of the Institut Henri Poincaré, is a review of the modern cosmological model. The introduction underlines the fundamental role played by general relativity in the genesis of the Big-Bang paradigm. The author carefully explains the hypotheses and principles on which this scenario relies. The first main part of the review, entitled 'The construction of the hot Big-Bang model', then describes the various stages in the development and confirmation of the Big-Bang paradigm from pure general relativity to the inclusion of matter and the description of large scale structures. It presents its two historical pillars, primordial nucleosynthesis and cosmic background radiation. The analysis leads to the conclusion that our universe started in a very dense and hot phase at thermal equilibrium, followed by cooling due to its expansion. Other critical aspects of the model include dark matter, dark energy and the formation of large-scale structure arising from tiny inhomogeneities in the early universe. This part ends with a detailed discussion of the status of the current standard cosmological model, namely the Λ-CDM (Lambda Cold Dark Matter) model, in which surprisingly only six basic parameters are needed to analyze and interpret all observational data. In the second half, entitled 'The primordial universe', the author turns to an in depth analysis of the current scenarios for the early history of our universe. The main concept is that of *inflation*, namely an early phase of accelerated expansion of space which can solve some of the main problems of the hot Big-Bang model. In particular, it smoothes out inhomogeneities, anisotropies and curvature and it causally connects distant regions of the universe as observed today. Beyond this, it predicts the existence of small inhomogeneities of quantum origin which explain the observed temperature anisotropies of the cosmic microwave background and are the seeds of the large scale structure of the universe. Although there is nowadays some consensus on the inflationary scenario, many details and questions remain to be understood, which the author discusses carefully, at some point meeting the unsolved problem of quantizing gravity. In this form, the contemporary cosmological models explain the origin and diversity of the atomic nuclei and of large scale structure and give some hints on the origin of matter. After having summarized the history and main successes of the Big-Bang theory, as well as its remaining problems and open questions, the review looks forward to the discovery of gravitational waves. The first observation of these waves in September 2015 opens a new observational window on our Universe.

The second article, entitled *The Planck Mission and the Cosmic Microwave Background*, is written by JEAN-LOUP PUGET, from the Institut d'Astrophysique Spatiale in Orsay. Launched on 14 May 2009, the *Planck* satellite was designed to orbit the L2 Lagrange point of the Sun-Earth system, and map the sky in nine frequencies using two state-of-the-art instruments: the Low Frequency Instrument (LFI), which includes three frequency bands in the range 30–70 GHz, and the High Frequency Instrument (HFI), which includes six frequency bands in the range 100–857 GHz. The latter's design was the most ambitious, with better sensitivity and angular resolution, but also more risky since it used cryogenically cooled bolometers at 100mK. J.-L. Puget, HFI's Principal Investigator, explains in great detail the historical path, both theoretical and experimental, that led to the *Planck* project and its specific and successful design. It was the third generation space mission after COBE and WMAP, with a stronger emphasis on the polarization of the CMB which is a much weaker signal than the temperature one (by a 50 to 100 factor), and seven of *Planck*'s nine frequency channels were equipped with polarization-sensitive detectors. In the end, *Planck* worked perfectly for 30 months, about twice the span originally required, and completed five full sky surveys with both instruments, before being turned off on 23rd October 2013.

The temperature and the polarization of the CMB shows minuscule fluctuations across the sky, which reflect the state of the cosmos at the time when light and matter parted company, 380 000 years after the Big Bang. This provides a powerful tool to estimate in a new and independent way parameters such as the age of the Universe, its rate of expansion (the Hubble constant) and its essential composition of normal matter, dark matter and dark energy. One main objective of the *Planck* mission was the most precise determination and constraint of the Λ-CDM cosmological model parameters, and the article explains the spectacular reduction of errors brought in by the *Planck* data release of 2015, by using jointly the CMB temperature and polarization spectra, the gravitational lensing signals, as well the final improvement brought in by observation data from galaxy surveys (such as Baryon Acoustic Oscillations). The pattern of acoustic oscillations in temperature and polarization power spectra implies an early Universe origin for the fluctuations, as in the inflationary framework of the Λ-CDM model. The primordial fluctuations are found to be Gaussian to an exceptional degree. There are no gravitational waves found at the 5% level, suggesting the energy scale of an inflationary epoch was below the Planck scale.

Planck's polarization data not only confirm and refine the details of the standard cosmological picture determined from the measurement of the CMB temperature fluctuations, but also help answer fundamental questions such as when the first stars began to shine. As their light interacted with gas in the Universe, atoms were gradually split back into electrons and protons during the 'reionization epoch'. The liberated electrons once again collided with the light from the CMB, albeit much less frequently in the expanded Universe, still leaving a telltale imprint on the polarization of the CMB. In 2016, a new analysis of the highly sensitive polarization measurements from *Planck*'s HFI, improving on earlier results from LFI in 2015, demonstrated that reionization was a very quick process, starting fairly late in cosmic history and having half-reionized the Universe by the time it was about 700 million years old.

The extraordinary wealth of high-quality data the mission has produced continued to be scientifically explored, see `https://www.cosmos.esa.int/web/Planck`. After completing a new processing of the data, the final *Planck* collaboration 'Legacy' results have been released on 17 July 2018, under the general title *Planck 2018 results*, and comprise twelve magnificent pieces; the first one, *Overview and the cosmological legacy of Planck*, is highly recommended and can be found at `https://www.aanda.org/articles/aa/full_html/2020/09/aa33880-18/aa33880-18.html`.

The importance of the results of the ESA Planck Scientific Collaboration for the Human scientific enterprise cannot be overstated: in addition to ten previous awards, including the 2017 Émilie du Chatelet Prize from Société française de Physique awarded to François Bouchet, IFI's Deputy PI, and in Year 2018 only, the Royal Astronomical Society presented the Planck Team with the 2018 Group Award, the same team and the PIs of LFI (Reno Mandolesi) and HFI (Jean-Loup Puget) were awarded the 2018 Gruber Cosmology Prize, Jean-Loup Puget received the 2018 Shaw Prize in Astronomy, and the 2018 Marcel Grossman Institutional Award went to the Planck Scientific Collaboration (ESA), represented by Jean-Loup Puget!

The third contribution, entitled *Massive Black Holes: Evidence, Demographics and Cosmic Evolution*, is due to REINHARD GENZEL, who leads the Max-Planck Institut für Extraterrestrische Physik (MPE) in Garching. He shared the 2020 Nobel Prize in Physics with ANDREA GHEZ from the University of California at Los Angeles *"for the discovery of a supermassive compact object at the centre of our galaxy"*, alongside ROGER PENROSE from the University of Oxford *"for the discovery that black hole formation is a robust prediction of the general theory of relativity"* (see `https://www.nobelprize.org/prizes/physics/2020/press-release/`). Obscured by thick clouds of absorbing dust, the closest supermassive black hole (MBH) to the Earth lies 26 000 light-years away at the center of our Galaxy, associated with the compact radio source SgrA*. It has a mass four million times that of the Sun, is surrounded by a small group of stars orbiting around it at speeds which can be relativistic. R. Genzel describes in exquisite experimental details the key difficulties of making very high angular resolution observations of the motion of the stars. The Schwarzschild radius of a 4 million solar mass black hole at the Galactic Center subtends a mere 10^{-5} arc-seconds, corresponding to about 2 cm at the distance of the Moon. The breakthrough came from the combination of Adaptive Optics techniques with advanced imaging and spectroscopic instruments, that allowed diffraction limited near-infrared spectroscopy and imaging astrometry with a precision of a few hundred micro-arc-seconds in the last decade. Using the Very Large Telescope (VLT) of the European Southern Observatory (ESO) in Chile, the group at MPE led by R. Genzel, and a group at UCLA led by A. M. Ghez with the Keck telescope in Hawai, independently found that the stellar velocities follow a 'Kepler' law as a function of distance from SgrA* and reach within the central light-month more than 10^3 km/s. By 2016, the two groups had determined individual orbits for more than 40 stars in the central light-month. The most spectacular of these stars, called S2, is in a highly elliptical orbit with a 16 year period. These orbits show that the gravitational potential indeed is that of a point mass centered on SgrA*, and concentrated well within the peri-approaches of the innermost stars, at 10-17 light-hours or 70 times the Earth orbit radius, and about 1000 times the event horizon of a 4 million solar mass BH. Combined with a proper motion limit of SgrA*, this lead to the inescapable conclusion that it can only be a MBH, eliminating astrophysical alternatives such as clusters of neutron stars, stellar black holes, or brown dwarfs.

The extreme gravitational field provided by the supermassive black hole makes it the perfect place to test Einstein's general theory of relativity. Finally, on July 26, 2018, it was announced that observations of the Galactic Center team at the MPE have for the first time revealed the relativistic effects on the motion of a star passing near the central MBH in the Milky Way. The exquisitely sensitive GRAVITY and SINFONI instruments in the VLT Interferometer allowed them to follow the star S2 as it approached the black hole during May 2018. At the closest point, the star was at a distance of only 17 light-hours, less than 20 billion kms from the MBH (120 times the Earth-Sun distance, and $1.5 \; 10^3$ times its Schwartzschild radius) and moving at a speed of almost 8000 km/s, about 2.5 % of the speed of light. The new measurements clearly reveal the gravitational redshift of the star's light, with a change in wavelength in precise agreement with Einstein's theory. The team used SINFONI to measure the velocity of S2 towards and away from Earth and the GRAVITY instrument to make extraordinarily precise measurements of the changing position of S2. With a gain of 15 in resolving power and precision, GRAVITY creates such sharp images that it can reveal the motion of the star from night to night as it 'grazes' the black hole — 26 000 light-years from Earth. This long-sought result represents the climax of a 26-year-long observation campaign using ESO's telescopes in Chile. (See http://www.mpe.mpg.de/6930756/news20180726.)

This volume ends with the thoroughly detailed contribution of ARNAUD CASSAN, a young astrophysicist at the Institut d'Astrophysique de Paris, entitled *New Worlds Ahead: The Discovery of Exoplanets*. This fascinating subject has developed at an ever accelerating rate since the identification of the first extrasolar planet in 1995 by Mayor and Queloz. We now have several hundred identified planetary systems with several thousand examples of exoplanets. The first part reviews all detection methods (pulsar timing, Doppler spectroscopy, transits, gravitational lensing, direct imaging and astrometry) with their respective advantages and limitations. The second part focuses on the astonishing diversity of planets and systems discovered, from the now famous hot Jupiters to super earths, from brown dwarves to planets orbiting pulsars. It implies a complete overhaul of our traditional Solar System paradigm.

The progress is so fast that it is difficult to catch up in this field. Since Cassan's paper, discoveries of Earth sized planets, some of them in the habitable zone of their parent star, have added to the excitation. Such Earth-like planets are as close as *Proxima Centauri b* (four light-years) or *Trappist 1* (*seven* earth-size rocky planets only about forty light years away, three of them possibly harboring liquid water). Science fiction now pales in light of reality. The first signs of exobiology may no longer be far off, as Cassan concludes that the field of exoplanet research has opened a door to the unknown.

This book, by the breadth of topics covered in both the theoretical description and present-day experimental study of the Universe, should be of broad interest to physicists, mathematicians, and historians of science. We further hope that the continued publication of this series of Proceedings will serve the scientific community, at both the professional and graduate levels. We thank the COMMISSARIAT À L'ÉNERGIE ATOMIQUE ET AUX ÉNERGIES ALTERNATIVES (Direction des Sciences de la Matière), the DANIEL IAGOLNITZER FOUNDATION, the ÉCOLE POLYTECHNIQUE, and the INSTITUT HENRI POINCARÉ for sponsoring this Seminar. Special thanks are due to Chantal DELONGEAS for the preparation of the manuscript.

Saclay & Orsay,
October 2020

BERTRAND DUPLANTIER
Université Paris-Saclay
CNRS, CEA
Institut de
Physique Théorique
91191, Gif-sur-Yvette, France
bertrand.duplantier@ipht.fr

VINCENT RIVASSEAU
Université Paris-Saclay
CNRS, Univ. Paris-Sud
Laboratoire de
Physique Théorique
91405, Orsay, France
rivass@th.u-psud.fr

Séminaire Poincaré XX

L'Univers

Samedi
21 novembre 2015

J.-P. Uzan : La théorie du Big Bang • *10h*

J.-L. Puget : Planck et le fond cosmologique • *11h*

R. Genzel : Massive Black Holes and Galaxy Evolution • *14h*

Y. Mellier : Les grandes structures de l'Univers • *15h*

A. Cassan : Les exoplanètes • *16h*

INSTITUT HENRI POINCARÉ • Amphi Hermite
11, rue Pierre et Marie Curie • 75005 Paris

www.bourbaphy.fr

FONDATION
IAGOLNITZER

ÉCOLE POLYTECHNIQUE • CPM • Image : NASA, ESA, N. Smith (University of California, Berkeley), and The Hubble Heritage Team (STScI/AURA)

The Universe, 1–72

Poincaré Seminar 2015

The Big-Bang Theory: Construction, Evolution and Status

Jean-Philippe Uzan

Abstract. Over the past century, rooted in the theory of general relativity, cosmology has developed a very successful physical model of the universe: the *big-bang model*. Its construction followed different stages to incorporate nuclear processes, the understanding of the matter present in the universe, a description of the early universe and of the large scale structure. This model has been confronted to a variety of observations that allow one to reconstruct its expansion history, its thermal history and the structuration of matter. Hence, what we refer to as the big-bang model today is radically different from what one may have had in mind a century ago. This construction changed our vision of the universe, both on observable scales and for the universe as a whole. It offers in particular physical models for the origins of the atomic nuclei, of matter and of the large scale structure. This text summarizes the main steps of the construction of the model, linking its main predictions to the observations that back them up. It also discusses its weaknesses, the open questions and problems, among which the need for a dark sector including dark matter and dark energy.

Keywords. Relativistic cosmology, physical cosmology, Big-Bang model.

1. Introduction

1.1. From General Relativity to cosmology

A cosmological model is a mathematical representation of our universe that is based on the laws of nature that have been validated locally in our Solar system and on their extrapolations (see Refs. [1, 2, 3] for a detailed discussion). It thus seats at the crossroad between theoretical physics and astronomy. Its basic enterprise is thus to use tested physical laws to understand the properties and evolution of our universe and of the matter and the astrophysical objects it contains.

Cosmology is however peculiar among sciences at least on two foundational aspects. The *uniqueness of the universe* limits the standard scientific method of comparing similar objects in order to find regularities and to test for reproductibility; indeed this limitation depends on the question that is asked. In particular, this will tend to blur many discussions on chance and necessity. Its *historical dimension* forces us to use abduction[1] together with deduction (and sometime induction) to reconstruct the most probable cosmological scenario[2]. One thus needs to reconstruct the conditions in the primordial universe to fit best what is observed at different epochs, given a set of physical laws. Again the distinction between laws and initial conditions may also be subtle.This means that cosmology also involves, whether we like it or not, some philosophical issues [4].

In particular, one carefully needs to distinguish *physical cosmology* from the *Cosmology* that aims to propose a global picture of the universe [1]. The former has tremendously progressed during the past decades, both from a theory and an observation point of view. Its goal is to relate the predictions of a physical theory of the universe to actual observations. It is thus mostly limited to our *observable universe*. The latter is aiming at answering broader questions on the universe as a whole, such as questions on origins or its finiteness but also on the apparent fine-tuning of the laws of nature for complexity to emerge or the universe to host a viable form of life. The boundary between these two approaches is ill-defined and moving, particularly when it comes to recent developments such as inflation or the multiverse debate. They are related to the two notions, the universe, i.e., the ensemble of all what exist, and our observable universe. Both have grown due to the progresses of our theories, that allow us to conceptualize new continents, and of the technologies, that have extended the domain of what we can observe and test.

Indeed the physical cosmology sets very strong passive constraints on Cosmology. It is then important to evaluate to which extent our observable universe is representative of the universe as a whole, a question whose answer depends drastically of what is meant by "universe as a whole". Both approaches are legitimate and the general public is mostly interested by the second. This is why we have the moral duty to state to which of those approaches we are referring to when we talk about cosmology.

While a topic of interest for many centuries – since any civilization needs to be structured by an anthropology and a cosmology, through mythology or science – we can safely declare [5] that scientific cosmology was born with Albert Einstein's general relativity a century ago. His theory of gravitation made the geometry of spacetime dynamical physical fields, $g_{\mu\nu}$, that need to be determined by solving equations known as Einstein field equations,

$$G_{\mu\nu}[g_{\alpha\beta}] = \frac{8\pi G}{c^4} T_{\mu\nu},$$ (1)

---

[1] Abduction is a form of inference which goes from an observation to a theory, ideally looking for the simplest and most likely explanation. In this reasoning, unlike with deduction, the premises do not guarantee the conclusion, so that it can be thought as "inference to the best explanation".
[2] A property cosmology shares with Darwinian evolution.

where the stress-energy tensor $T_{\mu\nu}$ characterizes the matter distribution. From this point of view, the cosmological question can be phrased as *What are the spacetime geometries and topologies that correspond to our universe?*

This already sets limitations on how well we can answer this question. First, from a pure mathematical perspective, the Einstein equations (1) cannot be solved in their full generality. They represent 10 coupled and non-linear partial differential equations for 10 functions of 4 variables and there is, at least for now, no general procedure to solve such a system. This concerns only the structure of the left-hand-side of Eq. (1). This explains the huge mathematical literature on the existence and stability of the solution of these equations.

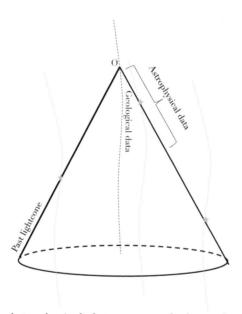

FIGURE 1. Astrophysical data are mostly located on our past lightcone and fade away with distance from us so that we have access to a portion of a 3-dimensional null hypersurface – an object can be observed only when its worldline (dashed lines) intersects our past lightcone. Some geological data can be extracted on our Solar system neighborhood. It is important to keep in mind that the interpretation of the observations is not independent of the spacetime structure, e.g., assigning distances. We are thus looking for compatibility between a universe model and these observations.

Another limitation arises from the source term in its right-hand-side. In order to solve these equations, one needs to have a good description of the matter content in the universe. Even with perfect data, the fact that (almost)[3] all the information we can extract from the universe is under the form of electromagnetic signal implies that observations are located on our past lightcone, that is on a 3-dimensional null hypersurface (see Fig. 1). It can be demonstrated (see, e.g., Ref. [6] and Ref. [7] for a concrete of 2 different cosmological spacetimes which enjoy the same lightcone observations) that the 4-dimensional metric cannot be reconstructed from this information alone. Indeed a further limitation arises from the fact that there is no such thing as perfect observations. Galaxy catalogs are limited in magnitude or redshift, evolution effects have to be taken into account, some components of matter (such a cold diffuse gas or dark matter) cannot be observed electromagnetically. We do not observe the whole matter distribution but rather classes of particular objects (stars, galaxies,. . .) and we need to deal with the variations in the properties of these individual objects and evolution effects. A difficult task is to quantify how the intrinsic properties of these objects influence our inference of the properties of the universe.

As a consequence, the cosmological question is replaced by a more modest question, the one of the construction of a good cosmological model, that can be phrased as *Can we determine metrics that are good approximations of our universe?* This means that we need to find a guide (such as symmetries) to exhibit some simplified solutions for the metrics that offer a good description on the universe on large scales. From a mathematical point of view, many such solutions are known [8]. Indeed our actual universe has no symmetry at all and these solutions have to be thought as a description of the universe smoothed on "some" scale, and we should not expect them to describe our spacetime from stellar scales to the Hubble scale.

The first relativistic cosmological model [9] was constructed by Einstein in 1917 and it can be considered as the birthdate of modern cosmology. As we shall see, most models rely on some hypotheses that are difficult to test on the size of the observable universe. The contemporary cosmological model is often referred to as the *big-bang* model and Section 2 describes its construction and structure. It is now complemented by a description of the early universe that we detail in Section 3, that offers both a model for the origin of the large scale structure but also a new picture of the universe as a whole. As any scientific model, it has to be compared to observation and a large activity in cosmology is devoted to understanding of how the universe looks to an observer inside this universe. Section 3 also sketches this theoretical activity and summarizes the observational landscape, unfortunately too wide to be described completely. This construction, while in agreement with all observation, relies on very simple, some may say crude, assumptions, which lead to the question *why does it works so well?* that we shall address in Section 4.

To summarize the main methodological peculiarities of cosmology, we have to keep in mind, (1) the uniqueness of the universe, (2) its historical dimension (hence the necessity of abduction), (3) the fact that we only reconstruct the most

---

[3]We also collect information in the Solar neighborhood and of high energy cosmic rays.

probable history (and then have to quantify its credence) that is backed up by the consistency of different facts (to be contrasted with a explanation designed to explain an isolated phenomena), (4) the need for a large extrapolation of the laws of nature, and (5) the existence of a (unspecified) smoothing scale.

Indeed, as for any model in physics, our model cannot explain its own ontology and opens limiting questions, such as its origin. The fact that we cannot answer them within the model is indeed no flaw and cannot be taken as an argument against the model. They just trivially show that it needs to be extended.

1.2. Hypotheses

The construction of a cosmological model depends on our knowledge of microphysics as well as on *a priori* hypotheses on the geometry of the spacetime describing our universe.

Theoretical physics describes the fundamental components of nature and their interactions. These laws can be probed locally by experiments. They need to be extrapolated to construct a cosmological model. On the one hand, any new idea or discovery will naturally call for an extension of the cosmological model (e.g., introducing massive neutrinos in cosmology is now mandatory). On the other hand, cosmology can help constraining extrapolations of the established reference theories to regimes that cannot be accessed locally. As explained above, the knowledge of the laws of microphysics is not sufficient to construct a physical representation of the universe. These are reasons for the need of extra-hypotheses, that we call cosmological hypotheses.

Astronomy confronts us with phenomena that we have to understand and explain consistently. This often requires the introduction of hypotheses beyond those of the physical theories in order to "*save the phenomena*" [10], as is actually the case with the dark sector of our cosmological model [11]. Needless to remind that even if a cosmological model is in agreement with all observations, whatever their accuracy, it does not prove that it is the "correct" model of the universe, in the sense that it is the correct cosmological extrapolation and solution of the local physical laws.

When confronted with an inconsistency in the model, one can either invoke the need for new physics, i.e., a modification of the laws of physics we have extrapolated in a regime outside of the domain of validity that has been established so far (e.g., large cosmological distance, low curvature regimes etc.), or have a more conservative attitude concerning fundamental physics and modify the cosmological hypotheses.

Let us start by reminding that the construction of any cosmological model relies on 4 main hypotheses (see Ref. [3] for a detailed description),

(H1) a theory of gravity,

(H2) a description of the matter contained in the universe and their non-gravitational interactions,

(H3) symmetry hypothesis,

(H4) a hypothesis on the global structure, i.e., the topology, of the universe.

These hypotheses are indeed not on the same footing since H1 and H2 refer to the local (fundamental) physical theories. These two hypotheses are however not sufficient to solve the field equations and we must make an assumption on the symmetries (H3) of the solutions describing our universe on large scales while H4 is an assumption on some global properties of these cosmological solutions, with same local geometry.

1.2.1. Gravity. Our reference cosmological model first assumes that gravity is well-described by general relativity (H1). This theory is well-tested on many scales and we have no reason to doubt it today [12, 13]. It follows that we shall assume that the gravitational sector is described by the Einstein-Hilbert action

$$S = \frac{1}{16\pi G} \int (R - 2\Lambda)\sqrt{-g}\mathrm{d}^4 x, \tag{2}$$

where a cosmological constant Λ has been included.

Indeed, we cannot exclude that it does not properly describe gravity on large scales and there exists a large variety of theories (e.g., scalar-tensor theories, massive gravity, etc.) that can significantly differ from general relativity in the early universe while being compatible with its predictions today. This means that we will have to design tests of general relativity on astrophysical scales [14, 15]. Indeed, from a theoretical point of view, we know that general relativity needs to be extended to a theory of quantum gravity. It is however difficult, on very general grounds, to determine if that would imply that there exist an "intermediate" theory of gravity that differs from general relativity on energy and distance scales that are relevant for the cosmological model. Indeed, there exist classes of theories, such as the scalar-tensor theories of gravity, that can be dynamically attracted [16] toward general cosmology during the cosmic evolution. Hence all the cosmological test of general relativity complement those on Solar system scales.

1.2.2. Non-gravitational sector. Einstein equivalence principle, as the heart of general relativity, also implies that the laws of non-gravitational physics validated locally can be extrapolated. In particular the constants of nature shall remain constant, a prediction that can also be tested on astrophysical scales [17, 18]. Our cosmological model assumes (H2) that the matter and non-gravitational interactions are described by the standard model of particle physics. As will be discussed later, but this is no breaking news, modern cosmology requires the universe to contain some dark matter (DM) and a non-vanishing cosmological constant (Λ). Their existence is inferred from cosmological observations assuming the validity of general relativity (e.g., flat rotation curves, large scale structure, dynamics of galaxy clusters for dark matter, accelerated cosmic expansion for the cosmological constant; see chapters 7 and 12 of Ref. [19]). Dark matter sets many questions on the standard model of particle physics and its possible extensions since the physical nature of this new field has to be determined and integrated consistently in the model. The cosmological constant problem is argued to be a sign of a multiverse, indeed a very controversial statement. If solved then one needs to infer some *dark energy* to be consistently included.

We thus assume that the action of the non-gravitational sector is of the form

$$S = \int \mathcal{L}(\psi, g_{\mu\nu})\sqrt{-g}\mathrm{d}^4x, \tag{3}$$

in which all the matter fields, ψ, are universally coupled to the spacetime metric.

Note that H2 also involves an extra-assumption since what will be required by the Einstein equations is the effective stress-energy tensor averaged on cosmological scales. It thus implicitly refers to a, usually not explicited, averaging procedure [20]. On large scale, matter is thus described by a mixture of pressureless matter ($P = 0$) and radiation ($P = \rho/3$).

1.2.3. Copernican principle. Let us now turn the cosmological hypotheses. In order to simplify the expected form of our world model, one first takes into account that observations, such as the cosmic microwave background or the distribution of galaxies, look isotropic around us. It follows that we may expect the metric to enjoy a local rotational symmetry and thus to be of the form

$$\mathrm{d}s^2 = -A^2(t,r)\mathrm{d}t^2 + B^2(t,r)\left[\mathrm{d}r^2 + R^2(t,r)\mathrm{d}\Omega^2\right]. \tag{4}$$

We are left with two possibilities. Either our universe is spherically symmetric and we are located close to its center or it has a higher symmetry and is also spatially homogeneous. Since we observe the universe from a single event, this cannot be decided observationally. It is thus postulated that we do not stand in a particular place of the universe, or equivalently that we can consider ourselves as a typical observer. This *Copernican principle* has strong implications since it implies that the universe is, at least on the size of the observable universe, spatially homogeneous and isotropic. Its validity can be tested [21] but no such test did actually exist before 2008. It is often distinguished from the *cosmological principle* that states that the universe is spatially homogeneous and isotropic. This latter statement makes an assumption on the universe on scales that cannot be observed [1]. From a technical point of view, it can be shown that it implies that the metric of the universe reduces to the Friedmann-Lemaître form (see, e.g., chapter 3 of Ref. [19])

$$\mathrm{d}s^2 = -\mathrm{d}t^2 + a^2(t)\left[\mathrm{d}\chi^2 + f_K^2(\chi)\mathrm{d}\Omega^2\right] \equiv g_{\mu\nu}\mathrm{d}x^\mu\mathrm{d}x^\nu, \tag{5}$$

where the scale factor a is a function of the cosmic time t. Because of the spatial homogeneity and isotropy, there exists a preferred slicing Σ_t of the spacetime that allows one to define this notion of cosmic time, i.e., Σ_t are constant t hypersurfaces. One can introduce the family of observers with worldlines orthogonal to Σ_t and actually show that they are following comoving geodesics. In terms of the tangent vector to their worldline, $u^\mu = \delta_0^\mu$, the metric (5) takes the form

$$\mathrm{d}s^2 = -(u_\mu\mathrm{d}x^\mu)^2 + (g_{\mu\nu} + u_\mu u_\nu)\mathrm{d}x^\mu\mathrm{d}x^\nu, \tag{6}$$

which clearly shows that the cosmic time t is the proper time measured by these fundamental observers. As a second consequence, this symmetry implies that the most general form of the stress-energy tensor is the one of a perfect fluid

$$T_{\mu\nu} = \rho u_\mu u_\nu + P(g_{\mu\nu} + u_\mu u_\nu), \tag{7}$$

with ρ and P, the energy density and isotropic pressure measured by the fundamental observers.

Now the 3-dimensional spatial hypersurfaces Σ_t are homogeneous and isotropic, which means that they are maximally symmetric. Their geometry can thus be only the one of either a locally spherical, Euclidean or hyperbolic space, depending on the sign of K. The form of $f_K(\chi)$ is thus given by

$$f_K(\chi) = \begin{cases} K^{-1/2} \sin\left(\sqrt{K}\chi\right), & K > 0, \\ \chi, & K = 0, \\ (-K)^{-1/2} \sinh\left(\sqrt{-K}\chi\right), & K < 0, \end{cases} \tag{8}$$

χ being the comoving radial coordinate. The causal structure of this class of space-times is discussed in details in Ref. [22].

1.2.4. Topology. The Copernican principle allowed us to fix the general form of the metric. Still, a freedom remains on the topology of the spatial sections [23]. It has to be compatible with the geometry.

In the case of a multiply connected universe, one can visualize space as the quotient X/Γ of a simply connected space X (which is just he Euclidean space E^3, the hypersphere S^3 or the 3-hyperboloid H^3, Γ being a discrete and fixed point free symmetry group of X. This holonomy group Γ changes the boundary conditions on all the functions defined on the spatial sections, which subsequently need to be Γ-periodic. Hence, the topology leaves the local physics unchanged while modifying the boundary conditions on all the fields. Given a field $\phi(\boldsymbol{x}, t)$ living on X, one can construct a field $\overline{\phi}(\boldsymbol{x}, t)$ leaving on X/Γ by projection as

$$\overline{\phi}(\boldsymbol{x}, t) = \frac{1}{|\Gamma|} \sum_{g \in \Gamma} \phi(g(\boldsymbol{x}), t), \tag{9}$$

since then, for all g, $\overline{\phi}(g(\boldsymbol{x}), t) = \overline{\phi}(\boldsymbol{x}, t)$. It follows that any Γ-periodic function of $L^2(X)$ can be identified to a function of $L^2(X/\Gamma)$.

In the standard model, it is assumed that the spatial sections are simply connected. The observational signature of a spatial topology decreases when the size of the universe becomes larger than the Hubble radius [24]. Its effects on the CMB anisotropy have been extensively studied [25] to conclude that a space with typical size larger than 20% of the Hubble radius today cannot be observationally distinguished from a infinite space [26]. From a theoretical point of view, inflation predicts that the universe is expected to be extremely larger than the Hubble radius (see below).

2. The construction of the hot big-bang model

The name of the theory, *big-bang*, was coined by one of its opponent, Fred Hoyle, during a BBC broadcast on March 28th 1948 for one was referred to the model of dynamical evolution before then. It became very media and one has to be aware that there exist many versions of this big-bang model and that it has tremendously evolved during the past century.

This section summarizes the main evolutions of the formulation of this model that is today in agreement with all observations, at the expense of introducing 6 free parameters. This version has now been adopted as the standard model for cosmology, used in the analysis and interpretation of all observational data.

2.1. General overview

The first era of relativistic cosmology, started in 1917 with the seminal paper by Einstein [9] in which he constructed, at the expense of the introduction of a cosmological constant, a static solution of its equations in which space enjoys the topology of a 3-sphere.

This paved the way to the derivations of exact solutions of the Einstein equations that offer possible world-models. Alexandr Friedmann and independently Georges Lemaître [27] developed the first dynamical models [28], hence discovering the cosmic expansion as a prediction of the equations of general relativity.

An important step was provided by Lemaître who connected the theoretical prediction of an expanding universe to observation by linking it to the redshifts of electromagnetic spectra, and thus of observed galaxies. This was later confirmed by the observations by Edwin Hubble [29] whose *Hubble law*, relating the recession velocity of a galaxy to its distance from us, confirms the cosmological expansion. The law of expansion derives from the Einstein equations and thus relates the cosmic expansion rate, H, to the matter content of the universe, offering the possibility to "weight the universe". This solution of a spatially homogeneous and isotropic expanding spacetime, referred to as Friedmann-Lemaître (FL) universe, serves as the reference background spacetime for most later developments of cosmology. It relies on the so-called Copernican principle stating that we do not seat in a particular place in the universe, and introduced by Einstein [9].

In a second era, starting in 1948, the properties of atomic and nuclear processes in an expanding universe were investigated (see, e.g., Ref. [30] for an early textbook). This allowed Ralph Alpher, Hans Bethe and George Gamow [31] to predict the existence and estimate the temperature of a cosmic microwave background (CMB) radiation and understand the synthesis of light nuclei, the big-bang nucleosynthesis (BBN), in the early universe. Both have led to theoretical developments compared successfully to observation. It was understood that the universe is filled with a thermal bath with a black-body spectrum, the temperature of which decreases with the expansion of the universe. The universe cools down and has a thermal history, and more important it was concluded that it emerges from a hot and dense phase at thermal equilibrium (see, e.g., Ref. [19] for the details).

This model has however several problems, such as the fact that the universe is spatially extremely close to Euclidean (*flatness problem*), the fact that it has an *initial spacelike singularity* (known as big-bang) and the fact that thermal equilibrium, homogeneity and isotropy are set as initial conditions and not explained (*horizon problem*). It is also too idealized since it describes no structure, i.e., does not account for the inhomogeneities of the matter, which is obviously distributed in galaxies, clusters and voids. The resolution of the naturalness of the initial conditions was solved by the postulate [32] of the existence of a primordial accelerated expansion phase, called *inflation*.

The third and fourth periods were triggered by an analysis of the growth of the density inhomogeneities by Lifshitz [33], opening the understanding of the evolution of the large scale structure of the universe, that is of the distribution of the galaxies in cluster, filaments and voids. Technically, it opens the way to the theory of cosmological perturbations [34, 35, 36] in which one considers the FL spacetime as a background spacetime the geometry and matter content of which are perturbed. The evolution of these perturbations can be derived from the Einstein equations. For the mechanism studied by Evgeny Lifshitz to be efficient, one needed initial density fluctuations large enough so that their growth in a expanding universe could lead to non-linear structure at least today. In particular, they cannot be thermal fluctuations. This motivated the question of the understanding of the origin and nature (amplitude, statistical distribution) of the initial density fluctuations, which turned out to be the second success of the inflationary theory which can be considered as the onset as the third era, the one of *primordial cosmology*. From a theory point of view, the origin of the density fluctuation lies into the quantum properties of matter [37]. From an observational point of view, the predictions of inflation can be related to the distribution of the large scale structure of the universe, in particular in the anisotropy of the temperature of the cosmic microwave background [38]. This makes the study of inflation an extremely interesting field since it requires to deal with both general relativity and quantum mechanics and has some observational imprints. It could thus be a window to a better understanding of quantum gravity.

The observational developments and the progresses in the theoretical of the understanding of the growth of the large scale structure led to the conclusion that

- there may exist a fairly substantial amount of non relativistic dark matter, or cold dark matter (hence the acronym CDM)
- there shall exist a non-vanishing cosmological constant (Λ).

This led to the formulation of the ΛCDM model [39] by Jeremy Ostriker and Paul Steinhardt in 1995. The community was reluctant to adopt this model until the results of the analysis of the Hubble diagram of type Ia supernovae in 1999 [40]. This ΛCDM model is in agreement with all the existing observations of the large survey (galaxy catalogs, CMB, weak lensing, Hubble diagram etc.) and its parameters are measured with increasing accuracy. This has opened the era of *observational cosmology* with the open question of the physical nature of the dark sector.

This is often advertised as *precision cosmology*, mostly because of the increase of the quality of the observations, which allow one to derive sharp constraints on the cosmological parameters. One has however to be aware, that these parameters are defined within a very specific model and require many theoretical developments (and approximations) to compare the predictions of the model to the data. Both set a limit a on the accuracy of the interpretation of the data; see, e.g., Refs. [41, 42] for an example of the influence of the small scale structure of the universe on the accuracy of the inference of the cosmological parameters. And indeed, measuring these parameters with a higher accuracy often does not shed more light on their physical nature.

The standard history of our universe, according to this model, is summarized in Fig. 2. Interestingly, the evolution of the universe and of its structures spans a period ranging from about 1 second after the big-bang to today. This is made possible by the fact that (1) the relevant microphysics is well-known in the range of energies reached by the thermal bath during that period (< 100 MeV typically) so that it involves no speculative physics, and (2) most of the observables can be described by a linear perturbation theory, which technically simplifies the analysis.

This description is in agreement with all observations performed so far (big-bang nucleosynthesis abundances, cosmic microwave background temperature and polarisation anisotropies, distribution of galaxies and galaxy clusters given by large catalogs and weak lensing observations, supernovae data and their implication for the Hubble diagram).

This short summary shows that today inflation is a cornerstone of the standard cosmological model and emphasizes its roles in the development and architecture of the model,

1. it was postulated in order to explain the required fine-tuning of the initial conditions of the hot big-bang model,
2. it provides a mechanism for the origin of the large scale structure,
3. it gives a new and unexpected vision of the universe on large scale,
4. it connects, in principle [43], cosmology to high energy physics.

This construction is the endpoint of about one century of theoretical and observational developments, that we will now detail.

2.2. Relativistic cosmology

2.2.1. Einstein static universe (1917).
The Einstein static universe is a static homogeneous and isotropic universe with compact spatial sections, hence enjoying the topology of a 3-sphere. It is thus characterized by 3 quantities: the radius of the 3-sphere, the matter density and the cosmological constant, that is mandatory to ensure staticity. They are related by

$$\Lambda = \frac{1}{R^2} = 4\pi G\rho. \tag{10}$$

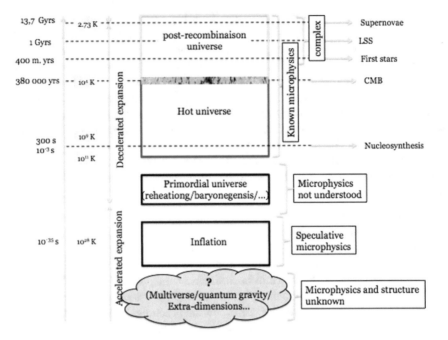

FIGURE 2. The standard history of our universe. The local universe provides observations on phenomena from big-bang nucleosynthesis to today spanning a range between 10^{-3} s to 13.7 Gyrs. One major transition is the equality which separates the universe in two era: a matter dominated era during which large scale structure can grow and a radiation dominated era during which the radiation pressure plays a central role, in particular in allowing for acoustic waves. Equality is followed by recombination, which can be observed through the CMB anisotropies. For temperatures larger that 10^{11} K, the microphysics is less understood and more speculative. Many phenomena such as baryogenesis and reheating still need to be understood in better details. The whole history of our universe appears as a parenthesis of decelerated expansion, during which complex structures can form, in between two periods of accelerated expansion, which do not allow for this complex structures to either appear or even survive. From Ref. [2].

so that the volume of the universe is $2\pi^2 R^3$.

2.2.2. de Sitter universe (1917). The same year, Wilhelm de Sitter constructed a solution that is not flat despite the fact that it contains no matter, but at the expense of having a cosmological constant. This solution was important in the discussion related to the Machs principle and today it plays an important role in inflation.

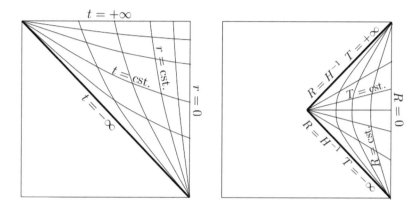

FIGURE 3. Penrose diagrams of de Sitter space in the flat (left) and static (right) slicings that each cover only part of the whole de Sitter space, and that are both geodesically incomplete. From Ref. [22].

The de Sitter spacetime is a maximally symmetric spacetime and has the structure of a 4-dimensional hyperboloid. It enjoys many slicings, only one of them being geodesically complete. These different representations, corresponding to different choices of the family of fundamental observers, are [22, 44]:

1. the spherical slicing in which the metric has a FL form (5) with $K = +1$ spatial sections and scale factor $a \propto \cosh Ht$ with $H = \sqrt{\Lambda/3}$ constant. It is the only geodesically complete representation;
2. the flat slicing in which the metric has a FL form (5) with Euclidean spatial sections, in which case $a \propto \exp Ht$;
3. the hyperbolic slicing in which the metric has a FL form (5) with $K = -1$ spatial sections and scale factor $a \propto \sinh Ht$;
4. the static slicing in which the metric takes the form $ds^2 = -(1-H^2R^2)dT^2 + dR^2/(1-H^2R^2) - r^2d\Omega^2$.

In each of the last three cases, the coordinate patch used covers only part of the de Sitter hyperboloid. Fig. 3 gives their Penrose diagrams to illustrate their causal structure (see Refs. [13, 22, 44] for the interpretation of these diagrams).

2.2.3. Dynamical models (1922–...) Starting from the previous assumptions, the spacetime geometry is described by the Friedmann-Lemaître metric (5). The Einstein equations with the stress-energy tensor (7) reduce to the Friedmann equations

$$H^2 = \frac{8\pi G}{3}\rho - \frac{K}{a^2} + \frac{\Lambda}{3}, \tag{11}$$

$$\frac{\ddot{a}}{a} = -\frac{4\pi G}{3}(\rho + 3P) + \frac{\Lambda}{3}, \tag{12}$$

together with the conservation equation $(\nabla_\mu T^{\mu\nu} = 0)$

$$\dot{\rho} + 3H(\rho + P) = 0. \tag{13}$$

This gives 2 independent equations for 3 variables (a, ρ, P) that requires the choice of an equation of state

$$P = w\rho \tag{14}$$

to be integrated. It is convenient to use the conformal time defined by $\mathrm{d}t = a(\eta)\mathrm{d}\eta$ and the normalized density parameters

$$\Omega_i = 8\pi G\rho_i/3H_0^2, \quad \Omega_\Lambda = \Lambda/3H_0^2, \quad \Omega_K = -K/a_0^2 H_0^2, \tag{15}$$

that satisfy, from Eq. (11), $\sum_i \Omega_i + \Omega_\Lambda + \Omega_K = 1$, so that the Friedmann equation takes the form

$$\frac{H^2}{H_0^2} = \sum_i \Omega_i(1+z)^{3(1+w_i)} + \Omega_K(1+z)^2 + \Omega_\Lambda \tag{16}$$

where the redshift z has been defined as $1 + z = a_0/a$. The Penrose diagram a FL spacetime with $\Lambda = 0$ is presented in Fig. 4 and the solutions of the Friedmann equations for different sets of cosmological parameters are depicted on Fig. 5.

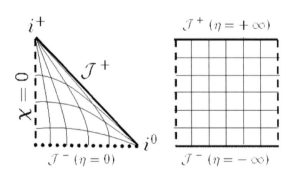

FIGURE 4. Conformal diagram of the Friedmann-Lemaître spacetimes with Euclidean spatial sections with $\Lambda = 0$ and $P > 0$ (left) and de Sitter space in the spherical slicing in which it is geodesically complete (right) – dashed line corresponds to $\chi = 0$ and $\chi = \pi$. From Ref. [19].

It is obvious from Eqs. (11-12) that the Einstein static solution and de Sitter solution are particular cases of this more general class of solutions. The general cosmic expansion history can be determined from these equations and are depicted in Fig. 5. A first property of the dynamics of these models can be obtained easily and without solving the Friedmann equations, by performing a Taylor expansion of $a(t)$ round $t = t_0$ (today). It gives $a(t) = a_0[1 + H_0(t - t_0) - \frac{1}{2}q_0 H_0^2(t - t_0)^2 + \ldots]$ where

$$H_0 = \left.\frac{\dot{a}}{a}\right|_{t=t_0} , \quad q_0 = -\left.\frac{\ddot{a}}{aH^2}\right|_{t=t_0} . \tag{17}$$

From the Friedmann equations, one deduces that $q_0 = \frac{1}{2}\sum(1 + 3w_i)\Omega_{i0} - \Omega_{\Lambda 0}$ so that the cosmic expansion cannot be accelerated unless there is a cosmological constant.

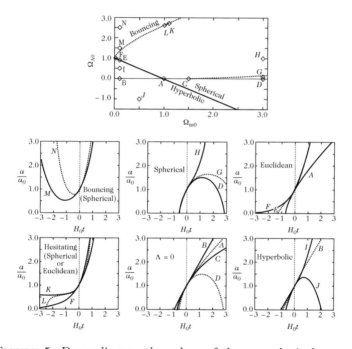

FIGURE 5. Depending on the values of the cosmological parameters (upper panel), the scale factor can have very different evolutions. It includes bouncing solutions with no big-bang (i.e., no initial singularity), hesitating universes (with a limiting case in which the universe is asymptotically initially Einstein static), collapsing universes. The expansion can be accelerating or decelerating. From Ref. [19].

In that description, the model can already be compared to some observations.

1. *Hubble law.* The first prediction of the model concerns the recession velocity of distant galaxies. Two galaxies with comoving coordinates 0 and $\boldsymbol{x}$ have a physical separation $\boldsymbol{r}(t) = a(t)\boldsymbol{x}$ so that their relative physical velocity is

$$\dot{\boldsymbol{r}} = H\boldsymbol{r},$$

as first established by Lemaître [27]. This gives the first observational pillar of the model and the order of magnitudes of the Hubble time and radius are obtained by expressing the current value of the Hubble parameter as $H_0 = 100\,h\,\mathrm{km}\cdot\mathrm{s}^{-1}\cdot\mathrm{Mpc}^{-1}$ with h typically of the of order 0.7 so that the present Hubble distance and time are

$$D_{\mathrm{H}_0} = 9.26\,h^{-1} \times 10^{25} \sim 3000\,h^{-1}\,\mathrm{Mpc}, \tag{18}$$

$$t_{\mathrm{H}_0} = 9.78\,h^{-1} \times 10^9\,\text{years}. \tag{19}$$

2. *Hubble diagram.* Hubble measurements [29] aimed at measuring independently distances and velocities. Today, we construct observationally a Hubble diagram that represents the distance in terms of the redshift. One needs to be careful when defining the distance since one needs to distinguish angular and luminosity distances. The luminosity distance can be shown to be given by

$$D_L(z) = a_0(1+z)f_K\left(\frac{1}{a_0 H_0}\int_0^z \frac{\mathrm{d}z'}{\mathbb{E}(z')}\right) \tag{20}$$

with $\mathbb{E} \equiv H/H_0$. The use of standard candels, such as type Ia supernovae allow to constrain the parameter of the models and measured the Hubble constant (see Fig. 6).

3. *Age of the universe.* Since $\mathrm{d}t = \mathrm{d}a/aH$, the dynamical age of the universe is given as

$$t_0 = t_{\mathrm{H}_0}\int_0^\infty \frac{\mathrm{d}z}{(1+z)\mathbb{E}(z)}. \tag{21}$$

A constraint on the cosmological parameters can also be obtained from the measurements of the age of the universe. A lower bound on this age can then be set that should be compatible with the dynamical age of the universe computed from the Friedmann equations, and it needs to be larger than the age of its oldest objects (see Chap. 4 of Ref. [19]).

During the development of this first formulation of the big-bang model, there is almost no discussion on the physical nature of the matter content (it is modeled as a matter & radiation fluid). The debate was primarily focused on the expansion, challenged by the steady state model. In the early stages (20ies), the main debate concerned the extragalactic nature of nebulae (the Shapley-Curtis debate). The model depends on only 4 independent parameters: the Hubble constant H_0 the density parameters for the matter and radiation, Ω_m and Ω_r, the spatial curvature, Ω_K, and the cosmological constant Ω_Λ so that the program of observational cosmology reduced mostly to measuring the mean density of the universe. From

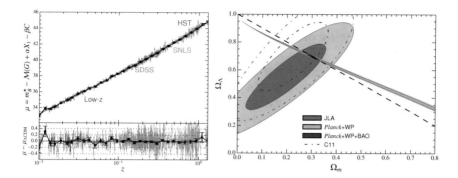

FIGURE 6. (left) A Hubble diagram obtained by the joint lightcurve analysis of 740 SNeIa from four different samples: Low-z, SDSS-II, SNLS3, and HST. The top panel depicts the Hubble diagram itself with the CDM best fit (black line); the bottom panel shows the residuals. (right) Constraints on the cosmological parameters obtained from this Hubble (together with other observations: CMB (green), and CMB+BAO (red). The dot-dashed contours corresponds to the constraints from earlier SN data. From Ref. [45].

these parameters one can infer an estimate of the age of the universe, which was an important input in the parallel debate on the legitimacy of a non-vanishing cosmological constant. The Copernican principle was also challenged by the derivation of non-isotropic solutions, such as the Bianchi [46] family, and non-inhomogeneous solutions, such as the Lemaître-Tolman-Bondi [47] spacetime.

2.3. The hot big-bang model

As challenged by Georges Lemaître [48] in 1931, *"une cosmogonie vraiment complète devrait expliquer les atomes comme les soleils."* This is mostly what the hot big-bang model will achieved.

This next evolution takes into account a better description of the matter content of the universe. Since for radiation $\rho \propto a^4$ while for pressureless matter $\rho \propto a^3$, it can be concluded that the universe was dominated by radiation. The density of radiation today is mostly determined by the temperature of the cosmic microwave background (see below) so that equality takes place at a redshift

$$z_{\mathrm{eq}} \simeq 3612\,\Theta_{2.7}^{-4}\left(\frac{\Omega_{\mathrm{m0}}h^2}{0.15}\right), \tag{22}$$

obtained by equating the matter and radiation energy densities and where $\Theta_{2.7} \equiv T_{\mathrm{CMB}}/2.725\,\mathrm{mK}$. Since the temperature scales as $(1+z)$, the temperature at which the matter and radiation densities were equal is $T_{\mathrm{eq}} = T_{\mathrm{CMB}}(1 + z_{\mathrm{eq}})$ which is of order

$$T_{\mathrm{eq}} \simeq 5.65\,\Theta_{2.7}^{-3}\,\Omega_{\mathrm{m0}}h^2\,\mathrm{eV}. \tag{23}$$

Above this energy, the matter content of the universe is in a very different form to that of today. As it expands, the photon bath cools down, which implies a *thermal history*. In particular,

- when the temperature T becomes larger than twice the rest mass m of a charged particle, the energy of a photon is large enough to produce particle-antiparticle pairs. Thus when $T \gg m_e$, both electrons and positrons were present in the universe, so that the particle content of the universe changes during its evolution, while it cools down;
- symmetries can be spontaneously broken;
- some interactions may be efficient only above a temperature, typically as long as the interaction rate Γ is larger than the Hubble expansion rate;
- the freeze-out of some interaction can lead to the existence of relic particles.

2.3.1. Equilibrium and beyond. Particles interaction are mainly characterized by a reaction rate Γ. If this reaction rate is much larger than the Hubble expansion rate, then it can maintain these particles in *thermodynamic equilibrium* at a temperature T. Particles can thus be treated as perfect Fermi-Dirac and Bose-Einstein gases with distribution[4]

$$F_i(E,T) = \frac{g_i}{(2\pi)^3} \frac{1}{\exp\left[(E - \mu_i)/T_i(t)\right] \pm 1} \equiv \frac{g_i}{(2\pi)^3} f_i(E,T), \qquad (24)$$

where g_i is the degeneracy factor, μ_i is the chemical potential and $E^2 = p^2 + m^2$. The normalisation of f_i is such that $f_i = 1$ for the maximum phase space density allowed by the Pauli principle for a fermion. T_i is the temperature associated with the given species and, by symmetry, it is a function of t alone, $T_i(t)$. Interacting species have the same temperature. Among these particles, the universe contains an electrodynamic radiation with black body spectrum (see below). Any species interacting with photons will hence have the same temperature as these photons as long as $\Gamma_i \gg H$. The photon temperature $T_\gamma = T$ will thus be called the *temperature of the universe*.

If the cross-section behaves as $\sigma \sim E^p \sim T^p$ (for instance $p = 2$ for electroweak interactions) then the reaction rate behaves as $\Gamma \sim n\sigma \sim T^{p+3}$; the Hubble parameter behaves as $H \sim T^2$ in the radiation period. Thus if $p + 1 > 0$, there will always be a temperature below which an interaction decouples while the universe cools down. The interaction is no longer efficient; it is then said to be *frozen*, and can no longer keep the equilibrium of the given species with the other components. This property is at the origin of the thermal history of the universe and of the existence of relics. This mechanism, during which an interaction can no longer maintain the equilibrium between various particles because of the cosmic expansion, is called *decoupling*. This criteria of comparing the reaction rate and the rate H is simple; it often gives a correct order of magnitude, but a more detailed description

---

[4]The distribution function depends *a priori* on $(\boldsymbol{x}, t)$ and $(\boldsymbol{p}, E)$ but the homogeneity hypothesis implies that it does not depend on $\boldsymbol{x}$ and isotropy implies that it is a function of $p^2 = \boldsymbol{p}^2$. Thus it follows from the cosmological principle that $f(\boldsymbol{x}, t, \boldsymbol{p}, E) = f(E, t) = f[E, T(t)]$.

of the decoupling should be based on a microscopic study of the evolution of the distribution function. As an example, consider Compton scattering. Its reaction rate, $\Gamma_{\text{Compton}} = n_e \sigma_T c$, is of order $\Gamma_{\text{Compton}} \sim 1.4 \times 10^{-3} H_0$ today, which means that, statistically, only one photon over 700 interacts with an electron in a Hubble time today. However, at a redshift $z \sim 10^3$, the electron density is 10^3 times larger and the Hubble expansion rate is of order $H \sim H_0 \sqrt{\Omega_{m0}(1+z)^3} \sim 2 \times 10^4 H_0$ so that $\Gamma_{\text{Compton}} \sim 80H$. This means that statistically at a redshift $z \sim 10^3$ a photon interacts with an electron about 80 times in a Hubble time. This illustrates that backward in times densities and temperature increase and interactions become more and more important.

As long as thermal equilibrium holds, one can define thermodynamical quantities such as the number density n, energy density ρ and pressure P as

$$n_i = \int F_i(\boldsymbol{p}, T)\mathrm{d}^3\boldsymbol{p}, \quad \rho_i = \int F_i(\boldsymbol{p}, T)E(\boldsymbol{p})\mathrm{d}^3\boldsymbol{p}, \quad P_i = \int F_i(\boldsymbol{p}, T)\frac{p^2}{3E}\mathrm{d}^3\boldsymbol{p}. \quad (25)$$

For ultra-relativistic particles $(m, \mu \ll T)$, the density at a given temperature T is then given by

$$\rho_{\text{r}}(T) = g_*(T)\left(\frac{\pi^2}{30}\right)T^4. \quad (26)$$

g_* represents the effective number of relativistic degrees of freedom at this temperature,

$$g_*(T) = \sum_{i=\text{bosons}} g_i \left(\frac{T_i}{T}\right)^4 + \frac{7}{8}\sum_{i=\text{fermions}} g_i \left(\frac{T_i}{T}\right)^4. \quad (27)$$

The factor $7/8$ arises from the difference between the Fermi and Bose distributions. In the radiation era, the Friedmann equation then takes the simple form

$$H^2 = \frac{8\pi G}{3}\left(\frac{\pi^2}{30}\right)g_* T^4. \quad (28)$$

Numerically, this amounts to

$$H(T) \cong 1.66 g_*^{1/2}\frac{T^2}{M_p}, \quad t(T) \cong 0.3 g_*^{-1/2}\frac{M_p}{T^2} \sim 2.42\, g_*^{-1/2}\left(\frac{T}{1\,\text{MeV}}\right)^{-2}\text{s}. \quad (29)$$

In order to follow the evolution of the matter content of the universe, it is convenient to have conserved quantities such as the entropy. It can be shown [19] to be defined as $S = sa^3$ in terms of the entropy density s as

$$s \equiv \frac{\rho + P - n\mu}{T}. \quad (30)$$

It satisfies $\mathrm{d}(sa^3) = -(\mu/T)\mathrm{d}(na^3)$ and his hence constant (i) as long as matter is neither destroyed nor created, since then na^3 is constant, or (ii) for non-degenerate relativistic matter, $\mu/T \ll 1$. In the cases relevant for cosmology, $\mathrm{d}(sa^3) = 0$. It can be expressed in terms of the temperature of the photon bath as

$$s = \frac{2\pi^2}{45} q_* T^3, \quad \text{with} \quad q_*(T) = \sum_{i=\text{bosons}} g_i \left(\frac{T_i}{T}\right)^3 + \frac{7}{8} \sum_{i=\text{fermions}} g_i \left(\frac{T_i}{T}\right)^3. \quad (31)$$

If all relativistic particles are at the same temperature, $T_i = T$, then $q_* = g_*$. Note also that $s = q_* \pi^4/45\zeta(3)n_\gamma \sim 1.8 q_* n_\gamma$, so that the photon number density gives a measure of the entropy.

The standard example of the use of entropy is the determination of the temperature of the cosmic neutrino background. Neutrinos are in equilibrium with the cosmic plasma as long as the reactions $\nu + \bar\nu \longleftrightarrow e + \bar e$ and $\nu + e \longleftrightarrow \nu + e$ can keep them coupled. Since neutrinos are not charged, they do not interact directly with photons. The cross-section of weak interactions is given by $\sigma \sim G_F^2 E^2 \propto G_F^2 T^2$ as long as the energy of the neutrinos is in the range $m_e \ll E \ll m_W$. The interaction rate is thus of the order of $\Gamma = n\langle\sigma v\rangle \simeq G_F^2 T^5$. We obtain that $\Gamma \simeq \left(\frac{T}{1\,\text{MeV}}\right)^3 H$. Thus close to $T_D \sim 1\,\text{MeV}$, neutrinos decouple from the cosmic plasma. For $T < T_D$, the neutrino temperature decreases as $T_\nu \propto a^{-1}$ and remains equal to the photon temperature.

Slightly after decoupling, the temperature becomes smaller than m_e. Between T_D and $T = m_e$ there are 4 fermionic states (e^-, e^+, each having $g_e = 2$) and 2 bosonic states (photons with $g_\gamma = 2$) in thermal equilibrium with the photons. We thus have that $q_\gamma(T > m_e) = \frac{11}{2}$ while for $T < m_e$ only the photons contribute to q_γ and hence $q_\gamma(T < m_e) = 2$. The conservation of entropy implies that after $\bar e - e$ annihilation, the temperatures of the neutrinos and the photons are related by

$$T_\gamma = \left(\frac{11}{4}\right)^{1/3} T_\nu. \quad (32)$$

Thus the temperature of the universe is increased by about 40% compared to the neutrino temperature during the annihilation. Since $n_\nu = (3/11)n_\gamma$, there must exist a cosmic background of neutrinos with a density of 112 neutrinos per cubic centimeter and per family, with a temperature of around 1.95 K today.

The evolution of any decoupled species can thus easily be described. However, the description of the decoupling, or of a freeze-out of an interaction, is a more complex problem which requires to go beyond the equilibrium description.

The evolution of the distribution function is obtained from the Boltzmann equation $L[f] = C[f]$, where C describes the collisions and $L = \mathrm{d}/\mathrm{d}s$ is the Liouville operator, with s the length along a worldline. The operator L is a function of eight variables taking the explicit form

$$L[f] = p^\alpha \frac{\partial}{\partial x^\alpha} - \Gamma^\alpha_{\beta\gamma} p^\beta p^\gamma \frac{\partial}{\partial p^\alpha}. \quad (33)$$

In a homogeneous and isotropic space-time, f is only function of the energy and time, $f(E,t)$, only so that

$$L[f] = E\frac{\partial f}{\partial t} - Hp^2\frac{\partial f}{\partial E}.$$ (34)

Integrating this equation with respect to the momentum $\boldsymbol{p}$, we obtain

$$\dot{n}_i + 3Hn_i = \mathcal{C}_i, \qquad \mathcal{C}_i = \frac{g_i}{(2\pi)^3}\int C\left[f_i(\boldsymbol{p_i},t)\right]\frac{\mathrm{d}^3\boldsymbol{p}_i}{E_i}.$$ (35)

The difficult part lies in the modelling and the evaluation of the collision term. In the simple of an interaction of the form $i + j \longleftrightarrow k + l$, the collision term can be decomposed as $\mathcal{C}_i = \mathcal{C}_{kl\to ij} - \mathcal{C}_{ij\to kl}$. There are thus 3 sources of evolution for the number density n_i, namely dilution $(3Hn_i)$, creation $(\mathcal{C}_i = \mathcal{C}_{kl\to ij})$ and destruction $(\mathcal{C}_i = \mathcal{C}_{ij\to kl})$.

To go further, one needs to specify the interaction. We will consider the nuclear interaction for BBN, the electromagnetic interaction for the CMB and particle annihilation for relics.

2.3.2. big-bang nucleosynthesis (BBN). big-bang nucleosynthesis describes the synthesis of light nuclei in the primordial universe. It is considered as the *second* pillar of the big-bang model. It is worth reminding that BBN has been essential in the past, first to estimate the baryonic density of the universe, and give an upper limit on the number of neutrino families, as was later confirmed from the measurement of the Z^0 width by LEP experiments at CERN.

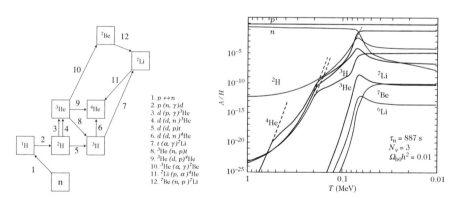

FIGURE 7. (left) The minimal 12 reactions network needed to compute the abundances up to lithium. (right) The evolution of the abundances of neutron, proton and the lightest elements as a function of temperature (i.e., time). Below 0.01 MeV, the abundances are frozen and can be considered as the primordial abundances. From Ref. [19].

Generalities on BBN.. The standard BBN scenario [19, 49, 50, 51] proceeds in three main steps:

1. for $T > 1$ MeV, ($t < 1$ s) a first stage during which the neutrons, protons, electrons, positrons an neutrinos are kept in statistical equilibrium by the (rapid) weak interaction

$$n \longleftrightarrow p + e^- + \bar{\nu}_e, \quad n + \nu_e \longleftrightarrow p + e^-, \quad n + e^+ \longleftrightarrow p + \bar{\nu}_e. \quad (36)$$

As long as statistical equilibrium holds, the neutron to proton ratio is

$$(n/p) = e^{-Q_{np}/k_B T} \quad (37)$$

where $Q_{np} \equiv (m_n - m_p)c^2 = 1.29$ MeV. The abundance of the other light elements is given by [19]

$$Y_A = g_A \left(\frac{\zeta(3)}{\sqrt{\pi}}\right)^{A-1} 2^{(3A-5)/2} A^{5/2} \left[\frac{k_B T}{m_N c^2}\right]^{3(A-1)/2} \eta^{A-1} Y_p^Z Y_n^{A-Z} e^{B_A/k_B T}, \quad (38)$$

where g_A is the number of degrees of freedom of the nucleus $_Z^A X$, m_N is the nucleon mass, η the baryon-photon ratio and $B_A \equiv (Z m_p + (A-Z) m_n - m_A)c^2$ the binding energy.

2. Around $T \sim 0.8$ MeV ($t \sim 2$ s), the weak interactions freeze out at a temperature T_f determined by the competition between the weak interaction rates and the expansion rate of the universe and thus roughly determined by $\Gamma_w(T_f) \sim H(T_f)$ that is

$$G_F^2 (k_B T_f)^5 \sim \sqrt{G N_*} (k_B T_f)^2 \quad (39)$$

where G_F is the Fermi constant and N_* the number of relativistic degrees of freedom at T_f. Below T_f, the number of neutrons and protons changes only from the neutron β-decay between T_f to $T_N \sim 0.1$ MeV when $p + n$ reactions proceed faster than their inverse dissociation.

3. For 0.05 MeV$< T < 0.6$ MeV (3 s $< t < 6$ min), the synthesis of light elements occurs only by two-body reactions. This requires the deuteron to be synthesized ($p + n \to D$) and the photon density must be low enough for the photo-dissociation to be negligible. This happens roughly when

$$\frac{n_d}{n_\gamma} \sim \eta^2 \exp(-B_D/T_N) \sim 1 \quad (40)$$

with $\eta \sim 3 \times 10^{-10}$. The abundance of ^{4}He by mass, Y_p, is then well estimated by

$$Y_p \simeq 2 \frac{(n/p)_N}{1 + (n/p)_N} \quad (41)$$

with

$$(n/p)_N = (n/p)_f \exp(-t_N/\tau_n) \quad (42)$$

with $t_N \propto G^{-1/2} T_N^{-2}$ and $\tau_n^{-1} = 1.636 G_F^2 (1 + 3g_A^2) m_e^5 / (2\pi^3)$, with $g_A \simeq 1.26$ being the axial/vector coupling of the nucleon.

4. The abundances of the light element abundances, Y_i, are then obtained by solving a series of nuclear reactions

$$\dot{Y}_i = J - \Gamma Y_i,$$

where J and Γ are time-dependent source and sink terms (see Fig. 7).

5. Today, BBN codes include up to 424 nuclear reaction network [52] with up-to-date nuclear physics. In standard BBN only D, ^{3}He, ^{4}He and ^{7}Li are significantly produced as well as traces of ^{6}Li, ^{9}Be, ^{10}B, ^{11}B and CNO. The most recent up-to-date predictions are discussed in Refs. [53, 54].

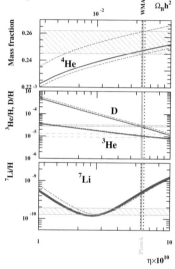

FIGURE 8. Abundances of ^{4}He D, ^{3}He and ^{7}Li (blue) as a function of the baryon over photon ratio (bottom) or baryonic density (top). The vertical areas correspond to the WMAP (dot, black) and Planck (solid, yellow) baryonic densities while the horizontal areas (green) represent the adopted observational abundances. The (red) dot–dashed lines correspond to the extreme values of the *effective* neutrino families coming from CMB Planck study, $N_{\rm eff} = (3.02, 3.70)$. From Ref. [54].

Observational status. These predictions need to be compared to the observation of the abundances of the different nuclei (Fig. 8), that we can summarized.

Deuterium is a very fragile isotope, easily destroyed after BBN. Its most primitive abundance is determined from the observation of clouds at high redshift, on the line of sight of distant quasars. Recently, more precise observations of damped Lyman-α systems at high redshift have lead to provide [55, 56] the mean value

$$\text{D/H} = (2.53 \pm 0.04) \times 10^{-5}. \tag{43}$$

After BBN, ^{4}He is still produced by stars, essentially during the main sequence phase. Its primitive abundance is deduced from observations in HII (ionized hydrogen) regions of compact blue galaxies. The primordial ^{4}He mass fraction, Y_p, is obtained from the extrapolation to zero metallicity but is affected by systematic uncertainties. Recently, [57] have determined that

$$Y_p = 0.2465 \pm 0.0097. \tag{44}$$

Contrary to ^{4}He, ^{3}He is both produced and destroyed in stars all along its galactic evolution, so that the evolution of its abundance as a function of time is subject to large uncertainties. Moreover, ^{3}He has been observed in our Galaxy [58], and one only gets a local constraint

$$^3\text{He/H} = (1.1 \pm 0.2) \times 10^{-5}. \tag{45}$$

Consequently, the baryometric status of ^{3}He is not firmly established [59].

Primitive lithium abundance is deduced from observations of low metallicity stars in the halo of our Galaxy where the lithium abundance is almost independent of metallicity, displaying the so-called Spite plateau [60]. This interpretation assumes that lithium has not been depleted at the surface of these stars, so that the presently observed abundance can be assumed to be equal to the primitive one. The small scatter of values around the Spite plateau is indeed an indication that depletion may not have been very efficient. However, there is a discrepancy between the value *i)* deduced from these observed spectroscopic abundances and *ii)* the BBN theoretical predictions assuming Ω_b is determined by the CMB observations. Many studies have been devoted to the resolution of this so-called *Lithium problem* and many possible "solutions", none fully satisfactory, have been proposed. For a detailed analysis see the proceedings of the meeting "Lithium in the cosmos" [61]. Note that the idea according to which introducing neutrons during BBN may solve the problem has today been shown [62] to be generically inconsistent with lithium and deuterium observations. Astronomical observations of these metal poor halo stars [63] have thus led to a relative primordial abundance of

$$\text{Li/H} = (1.58 \pm 0.31) \times 10^{-10}. \tag{46}$$

The origin of the light elements LiBeB, is a crossing point between optical and gamma spectroscopy, non thermal nucleosynthesis (via spallation with galactic cosmic ray), stellar evolution and big-bang nucleosynthesis. I shall not discuss them in details but just mention that typically, ^{6}Li/H$\sim 10^{-11}$. Beryllium is a fragile nucleus formed in the vicinity of Type II supernovae by non thermal process (spallation). The observations in metal poor stars provide a primitive abundance at very low metallicity of the order of Be/H = 3. $\times 10^{-14}$ at [Fe/H] = -4. This observation has to be compared to the typical primordial Be abundance, Be/H = 10^{-18}. Boron has two isotopes: ^{10}B and ^{11}B and is also synthesized by non thermal processes. The most recent observations give B/H = 1.7×10^{-12}, to be compared to the typical primordial B abundance B/H = 3×10^{-16}. For a general review of these light elements, see Ref. [64].

Finally, CNO elements are observed in the lowest metal poor stars (around [Fe/H]=-5). The observed abundance of CNO is typically [CNO/H]= -4, relatively to the solar abundance, i.e., primordial CNO/H< 10^{-7}. For a review see Ref. [65] and references therein.

Discussion. This shows that BBN is both a historical pillar and lively topic of the big-bang model. It allows one to test physics beyond the standard model. In particular, it can allow us to set strong constraints on the variation of fundamental constants [17, 66] and deviation from general relativity [67] As discussed it shows one of the major discrepancy of the model, namely the *Lithium problem*, that has for now no plausible explanation. Heavier elements such as CNO are now investigated in more details since they influence the evolution of Population III stars and invovled nuclear reactions the cross-section of which are badly known in the laboratory [54].

2.3.3. Cosmic microwave background. As long as the temperature of the universe remains large compared to the hydrogen ionisation energy, matter is ionized and photons are then strongly coupled to electrons through Compton scattering. At lower temperatures, the formation of neutral atoms is thermodynamically favored for matter. Compton scattering is then no longer efficient and radiation decouples from matter to give rise to a fossil radiation: the cosmic microwave background (CMB).

Recombination and decoupling. As long as the photoionisation reaction

$$p + e \longleftrightarrow H + \gamma \tag{47}$$

is able to maintain the equilibrium, the relative abundances of the electrons, protons and hydrogen will be fixed by the equation of chemical equilibrium. In this particular case, it is known as the Saha equation

$$\frac{X_e^2}{1 - X_e} = \left(\frac{m_e T}{2\pi}\right)^{3/2} \frac{e^{-E_I/T}}{n_b} \tag{48}$$

where $E_I = m_e + m_p - m_H = 13.6$ eV is the hydrogen ionisation energy and where one introduces the ionisation fraction

$$X_e = \frac{n_e}{n_p + n_H}, \tag{49}$$

where the denominator represents the total number of hydrogen nuclei, $n_b = n_p + n_H$ [$n_e = n_p = X_e n_b$, $n_H = (1 - X_e)n_b$] since electrical neutrality implies $n_e = n_p$.

The photon temperature is $T = 2.725(1 + z)$ K and the baryon density $n_b = \eta n_{\gamma 0}(1 + z)^3$ cm^{-3}. Notice first of all that when $T \sim E_I$, the right hand side of Eq. (48) is of order of 10^{15} so that $X_e(T \sim E_I) \sim 1$. Recombination only happens for $T \ll E_I$. X_e varies abruptly between $z = 1400$ and $z = 1200$ and the recombination can be estimated to occur at a temperature between 3100 and

3800 K. Note that Eq. (48) implies that

$$\frac{E_{\mathrm{I}}}{T} = \frac{3}{2}\ln\left(\frac{m_{\mathrm{e}}}{2\pi T}\right) - \ln\eta - \ln\left[\frac{2}{\pi^2}\zeta(3)\frac{X_{\mathrm{e}}^2}{1 - X_{\mathrm{e}}}\right],$$

which gives a rough estimates of the recombination temperature since the last term can be neglected; $T \sim 3500$ K.

The electron density varies quickly at the time of recombination, thus the reaction rate $\Gamma_{\mathrm{T}} = n_{\mathrm{e}}\sigma_{\mathrm{T}}$, with σ_{T} the Thompson scattering cross-section, drops off rapidly so that the reaction (47) freezes out and the photons decouple soon after. An estimate of the decoupling time, t_{dec}, can be obtained by the requirement $\Gamma_{\mathrm{T}}(t_{\mathrm{dec}}) = H(t_{\mathrm{dec}})$. It gives the decoupling redshift

$$(1 + z_{\mathrm{dec}})^{3/2} = \frac{280.01}{X_{\mathrm{e}}(\infty)}\left(\frac{\Omega_{\mathrm{b0}}h^2}{0.02}\right)^{-1}\left(\frac{\Omega_{\mathrm{m0}}h^2}{0.15}\right)^{1/2}\sqrt{1 + \frac{1 + z_{\mathrm{dec}}}{1 + z_{\mathrm{eq}}}}.\qquad(50)$$

$X_{\mathrm{e}}(\infty)$ is the residual electron fraction, once recombination has ended. We can estimate that $X_{\mathrm{e}}(\infty) \sim 7 \times 10^{-3}$.

This is indeed a simplified description since to describe the recombination, and to determine $X_{\mathrm{e}}(\infty)$, one should solve the Boltzmann equation and include the recombination of helium. A complete treatment, taking the hydrogen and helium contributions into account, is described in Ref. [19].

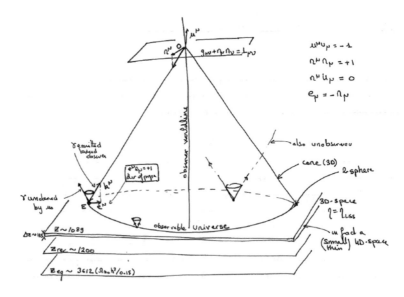

FIGURE 9. Spacetime diagram representing the equality hypersurface and the recombination followed by decoupling. Observed CMB photon are emitted on the intersection of this latter hypersurface and our past lightcone.

Last scattering surface. The optical depth can be computed once the ionisation fraction as a function of the redshift around decoupling is known,

$$\tau = \int n_e X_e \sigma_T d\chi. \tag{51}$$

It varies rapidly around $z \sim 1000$ so that the visibility function $g(z) = \exp(-\tau)$ $d\tau/dz$, which determines the probability for a photon to be scattered between z and $z + dz$, is a highly peaked function. Its maximum defines the decoupling time, $z_{dec} \simeq 1057.3$ (see Fig. 9). This redshift defines the time at which the CMB photons last scatter; the universe then rapidly becomes transparent and these photons can propagate freely in all directions. The universe is then embedded in this homogeneous and isotropic radiation. The instant when the photons last interact is a space-like hypersurface, called the *last scattering surface*. Some of these relic photons can be observed. They come from the intersection of the last scattering surface with our past light cone. It is thus a 2-dimensional sphere, centered around us, with comoving radius $\chi(z_{dec})$. This sphere is not infinitely thin. If we define its width as the zone where the visibility function is halved, we get $\Delta z_{dec} = 185.7$; see Fig. 9.

Properties of the cosmic microwave background. The temperature of the CMB, defined as the average of the temperature on the whole sky has been measured with precision by the FIRAS experiment on board of the COBE satellite [68]

$$T_0 = 2.725 \pm 0.001 \text{ K} \tag{52}$$

at 2σ. The observed spectrum is compatible with a black body spectrum

$$I(\nu) \propto \frac{\nu^3}{e^{\nu/T} - 1}. \tag{53}$$

The fact that this spectrum is so close to a black body proves that the fossil radiation could have been thermalized, mainly thanks to interactions with electrons. However, for redshifts lower than $z \sim 10^6$, the fossil radiation do not have time to be thermalized. Any energy injection at lower redshift would induce distortions in the Planck spectrum and can thus be constrained from these observations. It follows that

$$n_{\gamma 0} = 410.44 \, \Theta_{2.7}^3 \, \text{cm}^{-3} \quad \rho_{\gamma 0} = 4.6408 \times 10^{-34} \Theta_{2.7}^4 \, \text{g} \cdot \text{cm}^{-3} \tag{54}$$

and

$$\Omega_{\gamma 0} h^2 = 2.4697 \times 10^{-5} \Theta_{2.7}^4. \tag{55}$$

Note also that the temperature of the CMB can be measured at higher redshift to show that it scales as $(1 + z)$ hence offering an independent proof of the cosmic expansion (see Fig. 9).

Residual fluctuations. Once the monopole and dipole of the CMB have been removed, some temperature anisotropies remain, with relative amplitude $\sim 10^{-5}$, which correspond to temperature fluctuations of the order of 30 μK. These anisotropies correspond to anisotropies in the cosmic microwave background at

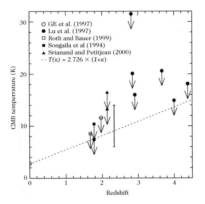

FIGURE 10. Measurements of the temperature of the CMB at different redshifts. The dashed line represents the prediction from the big-bang model. From Ref. [69].

the time of recombination and redshifted by the cosmic expansion. They can be decomposed on a basis of spherical harmonics as

$$\frac{\delta T}{T}(\vartheta, \varphi) = \sum_{\ell} \sum_{m=-\ell}^{m=+\ell} a_{\ell m} Y_{\ell m}(\vartheta, \varphi). \tag{56}$$

The angular power spectrum multipole $C_\ell = \langle |a_{lm}|^2 \rangle$ is the coefficient of the decomposition of the angular correlation function on Legendre polynomials.

This calls for an explanation to understand the origin of these fluctuations and their properties. They play a central role in modern cosmology that cannot be discussed in a smooth model of the universe.

2.3.4. Existence of relics. As the temperature of the universe drops, some interaction may freeze and let some thermal relics. This can be understood from Eq. (35) that can be rewritten as

$$\dot{n}_i + 3H n_i = -\langle \sigma v \rangle \left(n_i n_j - \frac{\bar{n}_i \bar{n}_j}{\bar{n}_k \bar{n}_l} n_k n_l \right), \tag{57}$$

where one sets $n_i = e^{\mu_i/T} \bar{n}_i$, $\bar{n}_i \equiv n_i[\mu_i = 0]$. $\langle \sigma v \rangle$ depends on the matrix elements of the reaction at hand.

As a simple example, consider a massive particle X in thermodynamic equilibrium with its anti-particle $\bar{X}$ for temperatures larger than its mass. Assuming this particle is stable, then its density can only be modified by annihilation or inverse annihilation $X + \bar{X} \longleftrightarrow l + \bar{l}$. If this particle had remained in thermodynamic equilibrium until today, its relic density, $n \propto (m/T)^{3/2} \exp(-m/T)$ would be completely negligible. The relic density of this massive particle, i.e., the residual

density once the annihilation is no longer efficient, will actually be more important since, in an expanding space, annihilation cannot keep particles in equilibrium
during the entire history of the universe. That particle is usually called a *relic*.

The equation (57) takes the integrated form

$$\dot{n}_X + 3H n_X = -\langle \sigma v \rangle \left(n_X^2 - \bar{n}_X^2 \right),$$ (58)

that can be integrated once $\langle \sigma v \rangle$ is described. In the case of cold relics, the decoupling occurs when the particle is non-relativistic. Assuming a form

$$\langle \sigma v \rangle = \sigma_0 f(x)$$ (59)

where $f(x)$ is a function of $x = m/T$ and $f(x) = x^{-n}$ one gets

$$\Omega_{X0} h^2 \simeq 0.31 \left[\frac{g_*(x_{\mathrm{f}})}{100} \right]^{-1/2} (n+1) x_{\mathrm{f}}^{n+1} \left(\frac{q_{*0}}{3.91} \right) \Theta_{2.7}^3 \left(\frac{\sigma_0}{10^{-38}\ \mathrm{cm}^2} \right)^{-1}$$ (60)

where x_{f} depends on the decoupling temperature.

This crude description shows that such relics can account for the dark matter.
Any freeze-out may lead to the existence of relics which should not contribute
too much to the matter budget of the universe, hence setting constraints on the
microphysics (such as, e.g., the problem with monopoles that can be produced in
the early universe).

2.3.5. Summary. The formulation of the hot big-bang model answered partly the
query by Lemaître. It offered a way to discuss the origin of the way the Mendeleev
table was populated and created a link with nuclear physics. It gives the historical
pillars of the model that can be connected to observations: BBN, CMB, expansion
of the universe. This demonstrates that the universe emerges from a hot and dense
phase at thermal equilibrium.

Its main problems are (1) the lithium problem, (2) the origin of the homogeneity of the universe (which is a question of the peculiarity of is initial conditions, (3)
the existence of an initial singularity and (4) the fact that it describes no structure,
contrary to an obvious observation.

2.4. Large scale structure and dark matter

The universe is indeed not smooth and a proper cosmological model needs to
address the origin and variety of the large scale structure (see Fig. 11). This means
that it needs to model the evolution of density fluctuations and find a scenario
for generating some initial density fluctuations. The first step is thus to develop a
theory of cosmological perturbation, which can deal only with a part of the objects
observed in the universe (see Fig. 11).

2.4.1. Perturbation theory. In order to describe the deviation from homogeneity,
one starts by considering the most general form of perturbed metric,

$$ds^2 = a^2(\eta) \left[-(1+2A)d\eta^2 + 2B_i dx^i d\eta + (\gamma_{ij} + h_{ij})dx^i dx^j \right],$$ (61)

where the small quantities A, B_i and h_{ij} are unknown functions of space and time
to be determined from the Einstein equations.

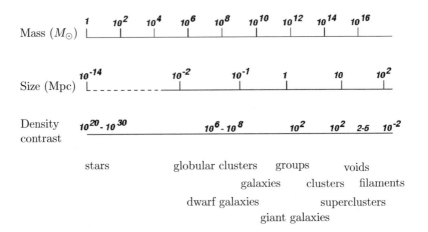

FIGURE 11. Scales associated with the different kinds of structures observed in the universe. Perturbation theory offers a tool to deal with the large scale structure only. (Courtesy of Yannick Mellier).

The strategy is then to write down the Einstein equations for this metric taking into account a perturbed stress-energy tensor. There are two important technical points in this program. We refer to Ref. [35] and to the chapters 5 and 8 of Ref. [19] or Ref. [13] for details.

- First, one can perform a Scalar-Vector-Tensor (SVT) decomposition of the perturbations and, at linear order, the modes will decouple. This decomposition is a generalization of the fact that any vector field can be decomposed as the sum of the gradient of a scalar and a divergenceless vector as

$$B^i = D^i B + \bar{B}^i, \quad h_{ij} = 2C\gamma_{ij} + 2D_i D_j E + 2D_{(i}\bar{E}_{j)} + 2\bar{E}_{ij}, \qquad (62)$$

with $D^i \bar{B}_i = 0$ and $D_i \bar{E}^{ij} = \bar{E}^i_i = 0$.

- Second, one has to carefully look at how the perturbation variables change under a gauge transformation, $x^\mu \to x^\mu - \xi^\mu$, where ξ^μ is decomposed into 2 scalar degrees of freedom and 2 vector degrees of freedom ($\bar{L}^i$, which is divergenceless $D_i \bar{L}^i = 0$) as

$$\xi^0 = T, \quad \xi^i = L^i = D^i L + \bar{L}^i. \qquad (63)$$

Under this change of coordinates, the metric and the scalar field transform as

$$g_{\mu\nu} \to g_{\mu\nu} + \mathcal{L}_\xi g_{\mu\nu}, \quad \varphi \to \varphi + \mathcal{L}_\xi \varphi.$$

In order to construct variables which remain unchanged under a gauge transformation, we introduce gauge invariant variables that "absorb" the components of ξ^μ. We thus get the 7 variables

$$\Psi \equiv -C - \mathcal{H}(B - E')$$
$$\Phi \equiv A + \mathcal{H}(B - E') + (B - E')', \tag{64}$$
$$\bar{\Phi}^i \equiv \bar{E}^{i\prime} - \bar{B}^i,$$

and it is obvious that the tensor modes $\bar{E}^{ij}$ were already gauge invariant.

The stress-energy tensor of matter now takes the general form $\bar{T}_{\mu\nu} + \delta T_{\mu\nu}$ with

$$\delta T_{\mu\nu} = (\delta\rho + \delta P)\bar{u}_\mu \bar{u}_\nu + \delta P \bar{g}_{\mu\nu} + 2(\rho + P)\bar{u}_{(\mu}\delta u_{\nu)} + P\delta g_{\mu\nu} + a^2 P \pi_{\mu\nu}, \tag{65}$$

where $u^\mu = \bar{u}^\mu + \delta u^\mu$ is the four-velocity of a comoving observer, satisfying $u_\mu u^\mu = -1$. The normalisation condition of $\bar{u}^\mu$ (to zeroth order) implies $\bar{u}^\mu = a^{-1}\delta_0^\mu$, $\bar{u}_\mu = -a\delta_\mu^0$, and since the norm of $\bar{u}^\mu + \delta u^\mu$ should also be equal to -1, we infer that $2\bar{u}^\mu \delta u_\mu + \delta g_{\mu\nu}\bar{u}^\mu\bar{u}^\nu = 0$ and thus that $\delta u_0 = -Aa$. We then write $\delta u^i \equiv v^i/a$ so that $\delta u^\mu = a^{-1}(-A, v^i)$ and decompose v_i into a scalar and a tensor part according to $v_i = D_i v + \bar{v}_i$. The anisotropic stress tensor $\pi_{\mu\nu}$ then consists only of a spatial part, which can be decomposed into scalar, vector and tensor parts as $\pi_{ij} = \Delta_{ij}\bar{p}i + D_{(i}\bar{p}_{j)} + \bar{p}_{ij}$, where the operator Δ_{ij} is defined as $\Delta_{ij} \equiv D_i D_j - \frac{1}{3}\gamma_{ij}\Delta$.

As for the metric perturbations, one can define gauge invariant quantities. Different choices are possible, and we define

$$\delta^N = \delta + \frac{\rho'}{\rho}(B - E'), \quad \delta^F = \delta - \frac{\rho'}{\rho}\frac{C}{\mathcal{H}}, \quad \delta^C = \delta + \frac{\rho'}{\rho}(v + B), \tag{66}$$
$$V = v + E', \quad \bar{V}_i = \bar{v}_i + \bar{B}_i. \tag{67}$$

The pressure perturbations are defined in an identical way. In all these relations, we recall that $\delta = \delta\rho/\rho$ is the density contrast.

The Einstein and conservation equations provide 3 sets of independent equations: one for the scalar variables (Φ, Ψ, δ, V), one for the two vector variables $(\bar{\Phi}_i$ and $\bar{V}_i)$ and one for the tensor mode $\bar{E}_{ij}$. To be solved, this system of equations needs to be closed by a description of the matter, that is by specifying an equation of state for the pressure and anisotropic stress.

We provide these equations for a single fluid but they can easily be derived for a mixture of fluids with non-gravitational integrations (see Chap. 5 of Ref. [19]).
Scalar modes. First of all, th Einstein equations provide two constraints,

$$(\Delta + 3K)\Psi = \frac{\kappa}{2}a^2\rho\delta^C, \quad \Psi - \Phi = \kappa a^2 P\bar{\pi}. \tag{68}$$

The first equation takes the classical form of the Poisson equation when expressed in terms of δ^C. The second equation tells us that the two gravitational potentials are equal if the scalar component of the anisotropic stress tensor vanishes.

The two other equations are

$$\Psi' + \mathcal{H}\Phi = -\frac{\kappa}{2}a^2\rho(1+w)V, \tag{69}$$

$$\Psi'' + 3\mathcal{H}(1+c_s^2)\Psi' + \left[2\mathcal{H}' + (\mathcal{H}^2 - K)(1+3c_s^2)\right]\Psi - c_s^2\Delta\Psi$$
$$= -9c_s^2\mathcal{H}^2\left(\mathcal{H}^2 + K\right)\bar{\pi}$$
$$- \left(\mathcal{H}^2 + 2\mathcal{H}' + K\right)\left[\frac{1}{2}\Gamma + (3\mathcal{H}^2 + 2\mathcal{H}')\bar{\pi} + \mathcal{H}\bar{\pi}' + \frac{1}{3}\Delta\bar{\pi}\right]. \tag{70}$$

The conservation of the stress-energy tensor gives a continuity and an Euler equation,

$$\left(\frac{\delta^{\mathrm{N}}}{1+w}\right)' = -(\Delta V - 3\Psi') - 3\mathcal{H}\frac{w}{1+w}\Gamma, \tag{71}$$

$$V' + \mathcal{H}(1 - 3c_s^2)V = -\Phi - \frac{c_s^2}{1+w}\delta^{\mathrm{N}} - \frac{w}{1+w}\left[\Gamma + \frac{2}{3}(\Delta + 3K)\bar{\pi}\right]. \tag{72}$$

These 2 equations are indeed redundant thanks to the Bianchi identities.

Vector modes. There are two Einstein equations for the vector modes,

$$(\Delta + 2K)\bar{\Phi}_i = -2\kappa\rho a^2(1+w)\bar{V}_i, \tag{73}$$

$$\bar{\Phi}'_i + 2\mathcal{H}\bar{\Phi}_i = \kappa P a^2 \bar{},\pi_i, \tag{74}$$

and here is a unique conservation equation,

$$\bar{V}'_i + \mathcal{H}\left(1 - 3c_s^2\right)\bar{V}_i = -\frac{1}{2}\frac{w}{1+w}(\Delta + 2K)\bar{\pi}_i. \tag{75}$$

It is clear from these equations that as long as there is no anisotropic stress source, the scalar modes decay away, $\bar{V}_i \propto a^{1-3c_s^2}$, $\bar{\Phi}_i \propto a^{-2}$. We shall thus not consider them so far even though they need to be included when topological defects, vector fields or magnetic fields are present.

Tensor modes. Their evolution is described by a single wave equation

$$\bar{E}''_{kl} + 2\mathcal{H}\bar{E}'_{kl} + (2K - \Delta)\bar{E}_{kl} = \kappa a^2 P\bar{\pi}_{kl}. \tag{76}$$

Fourier modes. As a consequence of the Copernican principle, these equations only involve the spatial Laplacian so they can easily be solved in Fourier space where they reduce to a set of coupled second order differential equations. (as a counter-example one may compare to the case of a Bianchi I universe in which modes do not decouple).

Two regimes need to be distinguished according to the value of the wavenumber k. For $k \ll \mathcal{H}$ the mode is said to be super-Hubble while for $k \gg \mathcal{H}$, is is said to be sub-Hubble.

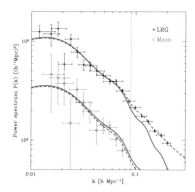

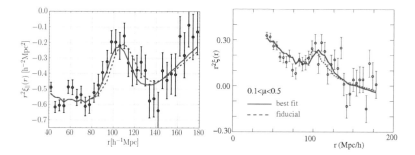

FIGURE 12. (left) Matter power spectra measured from the luminous red galaxy (LRG) sample and the main galaxy sample of the Sloan Digital Sky Survey (SDSS). Red solid lines indicate the predictions of the linear perturbation theory, while red dashed lines include nonlinear corrections. From Ref. [70]. (right) Two-point correlation function for objects aligned with the line of sight (top), or orthogonal to the line of sight (bottom), measured with the BOSS quasars ($2.1 \leq z \leq 3.5$) and the intergalactic medium traced by their Lyman-α forest, as function of comoving distance r. The effective redshift is $z = 2.34$ here. From Ref. [71].

2.4.2. Link to the observed universe. These equations set the stage for all the analysis of the large scale structure. In simple (academic) cases, they can be solved analytically, which brings some insight on the growth of the large scale structure. Let us just point out that for a pressureless fluid, neglecting K and considering sub-Hubble modes, the scalar equations reduce to their Newtonian counterpart

$$\Delta\Phi = 4\pi G\rho a^2 \delta, \quad \delta' = -\Delta V, \quad V' + \mathcal{H}V = -\Phi \tag{77}$$

where $\Psi = \Psi$ and the gauge can be forgotten. In this regime, we just get a close second order differential equations for δ: $\delta'' + 2\mathcal{H}\delta' - 4\pi G\rho\delta = 0$. It follows that the density perturbation can be split in terms of initial conditions and evolution as

$$\delta(\boldsymbol{k}, a) = D_+(\boldsymbol{k}, a)\delta_+(\boldsymbol{k}, a_i) + D_-(\boldsymbol{k}, a)\delta_-(\boldsymbol{k}, a_i). \tag{78}$$

It is then convenient to describe the evolution of the large scale structure in terms of a transfer function defined as

$$\Phi(k, a) = T(k, a)\Phi(k, a_i). \tag{79}$$

The observable modes today were initially super-Hubble and the theoretical models of the primordial universe (see below) allow us to predict the matter power spectrum of these modes, $P_\Phi(k, a_0) = P_\Phi(k, a_i)T^2(k, a_0)$, and $P_\delta(k, a_0) = P_\delta(k, a_i)T^2(k, a_0)\left[\frac{D_+(a_0)}{D_+(a_i)}\right]^2$.

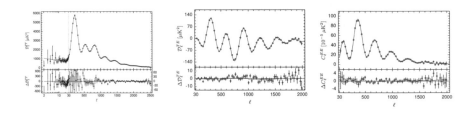

FIGURE 13. Angular power spectrum of the temperature (left), E-polarisation (right) anisotropies of the CMB, and their cross-correlation (middle) as measured by the Planck mission. From Ref. [72].

The general program can then be summarized as followed. One needs to determine the *transfer function*. It will depend on the species that are assumed to exist in the universe, on their non-gravitational interaction (e.g., the Compton scattering between photons and electrons). It thus depends on the cosmological parameters as well on physical parameters (e.g., the mass of the neutrinos). Note that one needs to go beyond a fluid description to correctly describe the photon and neutrinos, for which a Boltzmann equation needs to be derived.

Then, one needs to specify the *initial conditions*, either on an ad hoc way or by specifying a model of the primordial universe, such as inflation. This will provide the linear power spectrum. Independently of any model of the early universe, one can use the observation of the cosmic microwave background to constrain the initial power spectrum. According to the recent study by Planck [72], it is well-described by an almost scale invariant spectrum (see the text by J.-L. Puget in this volume).

This provides the statistical properties of the distribution of the large scale structures. They can then be compared to the observed distribution. The developments of large scale surveys is a growing field. Fig. 12 depicts the matter power spectra measured from the luminous red galaxy (LRG) sample and the main galaxy sample of the Sloan Digital Sky Survey [70]. In particular, it compares the predictions of the linear perturbation theory to non-linear corrections.

The comparison of the temperature anisotropies of the CMB to the distribution of the large scale structure confirms that the latter are formed through the gravitational collapse of initial density fluctuations of order of 10^{-5}. It also shows that if any, modifications compared to Einstein gravity have to be small. The baryon acoustic oscillation, that explains the peak structure of the CMB power spectrum corresponds to a rather large scale, which is weakly affected by the gravitational evolution of the universe between the epoch of recombination and today, contrary to small-scale inhomogeneities which tend to collapse and lose information about their initial conditions. As a consequence, this correlation within the distribution of baryonic matter in the universe has survived, only growing with cosmic expansion. They have been observed (see Fig. 12) and demonstrate the consistency of this picture. Indeed, the comparison of these different observables allow one to sharpen the constraints on the transfer function and thus on the cosmological parameters.

One then needs to go beyond the linear regime. One can either work out the weakly non-linear regime perturbatively or resort on numerical simulation. The first approach can be performed either in a full-relativistic set-up or in the Newtonian regime while numerical simulations remain Newtonian. Both methods have their own limitations. In particular reaching sub-percent precision in small scales (typically $1 - 10h$ Mpc) requires to control the validity of higher perturbation expansion. It was estimated [103] that the lowest mode k at which linear perturbation theory deviates from a reference spectrum is 0.03 at $z = 0$ and reaches 0.09 ate $z = 1.5$ while with 1-loop (resp. 2-loop) correction is 0.08 (resp. 0.04) at $z = 0$ and 0.14 (resp. 0.20) at $z = 1.5$. Various alternative techniques, e.g., based on the renormalisation group, have been developed in order to understand the effect of the small scales on the evolution of perturbations, hence leading to the idea that one may be able to construct an effective field theory of the large scale structure [74]. Note also that most theoretical results are based on the idealized system of a collisionless cold matter fluid (described by point particles) that interacts only gravitationally. It is well-suited for cold dark matter, the mean free path of which is small but requires more attention when it comes to hot species, such as neutrinos which remain a challenge. And indeed, pressure has to be taken into account when one introduces baryons. We can conclude that we have a good understanding of the physics at work but that high precision and realistic predictions still require progresses.

2.4.3. Dark matter and Λ. The analysis of the large scale structure has led to a series of indications for the need of a dark sector composed of two components:

- *dark matter* that can be assimilated to a non-interacting and non-baryonic dust component;
- *dark energy* that can be modeled by a smooth fluid with a negative pressure, the best candidate both from a phenomenological and observational point of view being the cosmological constant.

The evidence for the existence of dark matter can be traced back to the analysis by Zwicky of the dynamics of the Coma cluster [75] in 1933, and then to the first rotation curves for the Andromeda nebula [76] in 1939. The ideas that galaxies and clusters have a dark matter halo was not widely considered before the seventies, the change mostly triggered by Refs. [77, 78] and the arguments that massive halos are required to stabilize spiral galaxy disks [79]. It was then followed by a formulation of the basis of a galaxy formation scenario [80] in which a hierarchically merging population of dark matter halos provide the gravitational potential wells in which the intergalactic gas can cool and condense to form galaxies. In that model, the evolution of the distribution of the halo is described by the Press-Schechter analytic model [81]. Note also that combining the growth rate of the perturbation, as obtained by Lifschitz [33], that shows that is scales as a, and the observational amplitude of density fluctuations from the CMB, $\delta\rho_b/\rho_b \sim 10^{-5}$, the perturbation cannot become non-linear for a decoupling at $z \sim 10^3$. This gives another indication for the need of dark matter, to form potential wells before decoupling. Many scenarios of structure formation were then investigated, mostly on the basis of numerical simulations, distinguishing between warm (i.e., relativistic) and cold (i.e., non-relativistic) dark matter.

Among the potential candidates, a massive neutrino was the one to be known experimentally but it was shown not to offer a good cosmological solution. Today, it is important to remind that the need for dark matter in our cosmological model is one of the strong phenomenological arguments for the need to physics beyond the standard model. Many candidates such as axions or supersymmetric particles have been considered, and cosmology offers an important observational constraint on these models.

To summarize the proofs for the existence of dark matter comes from many observations: (1) the dynamics of gravitational bound objects (such as galaxies and clusters), (2) weak lensing effects, (3) the fact that baryonic matter interacts with radiation while dark matter does not (related to the amplitude of the peak structure of the CMB or the bullet cluster observations). Indeed several model tend to explain these observation by a modification of general relativity, usually in a low acceleration regime. It is important to say that these models also need to invoke new degrees of freedom that needs to be coupled to the standard matter. The main difference arise from the fact that in dark matter models, the energy density of the new field is an extra-source to the gravitational potential while in modified general relativity models, a new long-range interaction modifies the gravitational force. The current status is that none of these latter models can explain dark matter on all scales from galaxies to the CMB.

The observation of type Ia supernovae (see Fig. 6) shows that the cosmic expansion is accelerating today. From the Friedmann equations (12) it can happen only if the matter content of the universe is dominated by a component such that $\rho + 3P < 0$, which has been named *dark energy*. The natural candidate is a cosmological constant, and no observation shows any deviation from this hypothesis (e.g., a variation with redshift of the equation of state of this component). The main problem arises from the amplitude of the associated energy density $\rho_\Lambda = \Lambda / 8\pi G$ compared to what is inferred for the vacuum energy from quantum field theory in curved spacetime. This has split the approaches in two very different avenues. One the one hand, one can accept a cosmological constant and argue that the resolution of the cosmological constant lies in a multiverse hypothesis. On the other hand, one assumes the problem is solved (but no concrete way of the solution does actually exist) and invoke a new degrees of freedom. They can be either geometrical (i.e., by discussing that we have been fooled by assuming the validity of the FL geometry) or physical (i.e., new fields in the theory). In that latter case, as for dark matter, they can be associated to a modification of general relativity on large scale or not. It is also important to state that all of these models are phenomenological and of importance to discuss what cosmological can detect or constrain but none of these models is on safe theoretical grounds.

Today, we shall face that no deviation from general relativity+Λ have been observationally detected and this is the best phenomenological model so far, hence that shall be considered as the reference model. Indeed, we know that general relativity needs to be extended to a quantum description, but it is hard to connect this to the required modification of general relativity to account for the acceleration of the universe, in terms of energy scales, and on the fact that it shall also solve the cosmological constant problem. See, e.g., Refs. [3, 11, 19] for further discussions and references.

2.5. Modelling observations

Most cosmological measurements rely, so far, on the observation of distant light sources, such as galaxies, supernovae, or quasars. Interpreting these observations requires to know the optical properties of the universe. In particular, the construction of a distance scale has been a long standing problem in astronomy.

2.5.1. Light propagation and distances.
The interpretation of all cosmological observations relies on two equations that describe the propagation of a light ray and the evolution of the geodesic bundle, the first being more fundamental since it also applies in case of string lensing.

Fundamental equations. The first is the geodesic equation for the tangent vector, $k^\mu = \mathrm{d}x^\mu / \mathrm{d}\lambda$ to null geodesics $x^\mu(\lambda)$,

$$k^\mu \nabla_\mu k^\nu = 0, \quad k^\mu k_\mu = 0 \tag{80}$$

that derives from the Maxwell equation, $\nabla_\mu F^{\mu\nu} = 0$ in the eikonal approximation. Its integration allows one to determine our past lightcone, that is to obtain the position of any object as its position on the sky $\{\theta(\lambda), \varphi(\lambda)\}$ and its redshift $z(\lambda)$, from which its distance needs to be determined, which means it is the byproduct of a model and not a direct observable.

In order to study object of finite size, one shall also describe the relative behavior of two neighboring geodesics, that is a bundle of geodesics, $x^\mu(\cdot, y^a)$ and $x^\mu(\cdot, y^a + \delta y^a)$, where $\xi^\mu = (\partial x^\mu / \partial y^a) \delta y^a$ is the connecting vector between two geodesics. If the origin $v = 0$ of the affine parametrization of all rays is taken at O, then $k^\mu \xi_\mu = 0$ and then, the evolution of ξ^μ along the beam is governed by the geodesic deviation equation

$$k^\alpha k^\beta \nabla_\alpha \nabla_\beta \xi^\mu = R^\mu{}_{\nu\alpha\beta} k^\nu k^\alpha \xi^\beta, \tag{81}$$

where $R^\mu{}_{\nu\alpha\beta}$ is the Riemann tensor.

Consider an observer, with four-velocity u^μ ($u_\mu u^\mu = -1$), who crosses the light beam. The tangent vector to the beam can always be decomposed as $k^\mu = -\omega(u^\mu + d^\mu)$ where where $\omega = u_\mu k^\mu$. The redshift z in the observer's rest-frame is thus given by

$$1 + z = \frac{\nu_s}{\nu_o} = \frac{u_s^\mu k_\mu(v_s)}{u_o^\mu k_\mu(v_o)}. \tag{82}$$

These two equations allow to construct all observables. The difficult point lies in the fact that one needs to express them consistently in the cosmological framework, which in many cases is a difficult task.

Sachs formalism. In order to measure shapes, the observer needs to define a screen (i.e., a 2-dimensional space on which the cross-section of the beam is projected) and spanned by 2 vectors s_A^μ. Once projected on this basis, the geodesic equation reduces to the Sachs equation

$$\frac{d^2 \xi_A}{dv^2} = \mathcal{R}_{AB} \xi^B, \tag{83}$$

where $\xi_A = s_A^\mu \xi_\mu$ and $\mathcal{R}_{AB} = R_{\mu\nu\alpha\beta} k^\nu k^\alpha s_A^\mu s_B^\beta$ are the screen-projected connecting vector and Riemann tensor, usually called the *optical tidal matrix*. The properties of the Riemann tensor imply that this matrix is symmetric, $\mathcal{R}_{AB} = \mathcal{R}_{BA}$. Introducing the Jacobi matrix as

$$\xi^A(v) = \mathcal{D}_B^A(v) \dot{\xi}^B(0). \tag{84}$$

that is as relating the physical separation $\xi^A(v)$ of two neighbouring rays of a beam at v to their observed separation $\dot{\xi}^A(0)$, the Sachs equation takes the general form

$$\ddot{\mathcal{D}}_B^A = \mathcal{R}_C^A \mathcal{D}_B^C, \quad \mathcal{D}_B^A(0) = \delta_B^A, \quad \dot{\mathcal{D}}_B^A(0) = 0. \tag{85}$$

Distances. First of all, up to frequency factor ω_o fixed to 1 here, the determinant of the Jacobi matrix is related to the angular diameter distance as

$$D_A^2(v) = \frac{\text{area of the source at } v}{\text{observed angular size}} = \det \mathcal{D}_B^A(v). \tag{86}$$

The luminosity distance is then defined as

$$\phi_{\text{obs}} = \frac{L_{\text{source}}(\chi)}{4\pi D_{\text{L}}^2}, \tag{87}$$

that relates the intrinsic luminosity of a source to the observed flux.

Interestingly, one can demonstrate a general theorem, that holds as long as the photon number is conserved, between the luminosity and angular distances, known as the distance duality relation

$$D_L = (1+z)^2 D_A(z). \tag{88}$$

Such a relation can be tested observationally [82] and allows one to rule out some dark energy models [83].

2.5.2. Background universe. In a FL universe, these distances can be determined analytically and are given by Eq. (20). This sets the basis of the interpretation of the Hubble diagram; see Fig. 6. At small redshifts, it can be expanded to get

$$D_A(z) = D_{\text{H}_0}\left[1 - \frac{1}{2}(q_0 + 3)z\right]z + \mathcal{O}(z^3), \tag{89}$$

and

$$D_{\text{L}}(z) = D_{\text{H}_0}\left[1 - \frac{1}{2}(q_0 - 1)z\right]z + \mathcal{O}(z^3). \tag{90}$$

Locally ($z \ll 1$) all the distances reduce to $D_{\text{H}_0}z$ and the deviations only depend on the value of the deceleration parameter.

2.5.3. Perturbation theory.
Generalities. When dealing with perturbation, one needs to solve the geodesic and Sachs equation at least at linear order in perturbation.

Each observable needs to be related to the expression of the gauge invariant perturbation variable. This requires to define clearly what is actually observed. At linear order, these computations are well-established but beyond linear order there are still some debates.

I just give a short overlook of two important examples: weak-lensing and the CMB anisotropies, in order to relate to the contributions by Yannick Mellier and Jean-Loup Puget. A full derivation can be obtained in chapters 6 and 7 of Ref. [19].

Weak lensing. Usually, when dealing with weak lensing, treated as perturbation with respect to a FL spacetime, one uses that at background level both the shear and rotation vanish so that $\bar{\mathcal{D}}_B^C = \bar{D}_A\,\delta_B^C$. At linear order, it can be expanded to get the definition of the *amplification matrix* as

$$\mathcal{D}_B^A = \mathcal{A}_C^A\bar{\mathcal{D}}_B^C + \mathcal{O}(2) \tag{91}$$

with

$$\mathcal{A}_C^A = \begin{pmatrix} 1 - \kappa - \gamma_1 & \gamma_2 + \psi \\ \gamma_2 - \psi & 1 - \kappa + \gamma_1 \end{pmatrix}, \tag{92}$$

where the *convergence* is defined as $\kappa = 1 - \frac{1}{2}\text{Tr}\mathcal{A} = \frac{D_A - \bar{D}_A}{\bar{D}_A} + \mathcal{O}(2)$.

By expanding the Jacobi equation (85) to linear order, the amplification matrix in a cosmological space-time can thus be expressed in terms of the gravitational potential as $\mathcal{A}_{ab} = \delta_{ab} - \partial_{ab}\psi(\boldsymbol{\theta}, \chi)$ with the lensing potential

$$\psi(\boldsymbol{\theta}, \chi) \equiv \frac{1}{c^2} \int_0^\chi \frac{f_K(\chi')f_K(\chi - \chi')}{f_K(\chi)} (\Phi + \Psi) [f_K(\chi')\boldsymbol{\theta}, \chi']\mathrm{d}\chi'. \qquad (93)$$

It follows that

$$\kappa(\boldsymbol{\theta}, \chi) = \frac{1}{2} (\psi_{,11} + \psi_{22}),$$

$$\gamma_1(\boldsymbol{\theta}, \chi) = \frac{1}{2} (\psi_{,11} - \psi_{22}), \qquad (94)$$

$$\gamma_2(\boldsymbol{\theta}, \chi) = \psi_{,12}.$$

In order to be related to observation, one shall integrate over the source distribution, $p_z(z)\mathrm{d}z = p_\chi(\chi)\mathrm{d}\chi$, and the effective convergence is obtained by weighting the convergence (94) with the source distribution as

$$\kappa(\boldsymbol{\theta}) = \int p_\chi(\chi)\kappa(\boldsymbol{\theta}, \chi)\mathrm{d}\chi.$$

Splitting the perturbations as a transfer function, obtained by solving the perturbation equations, and initial conditions described in terms of an initial power spectrum, one can then work out the angular power spectrum of these observables (after a decomposition in spherical harmonics). It can be demonstrated that $P_\kappa = P_\gamma$.

It follows that the key in the measurement of the cosmic shear relies on the measurement of the shape of background galaxies and on their ellipticity (see Refs. [84, 85, 86] for details and problems). If the image of such a galaxy is represented by an ellipticity $\varepsilon = \varepsilon_1 + i\varepsilon_2 = (1 - r)/(1 + r) \exp(2i\phi)$, $r = b/a$ being the ratio between the major and minor axes, then

$$\langle \varepsilon \rangle = \left\langle \frac{\gamma}{1 - \kappa} \right\rangle.$$

In the regime of weak distortion ($\kappa \ll 1$), the ellipticities give access to the shear.

This techniques has witnessed tremendous progresses in the past 15 years. They are described in the contribution by Yannick Mellier. It is important to stress that weak lensing gives a way to directly measure the distribution of the gravitational potential. Combined with the observation of the matter distribution and of velocity fields, it gives a way to test the prediction of general relativity, as first pointed out in Ref. [15] (see also Ref. [14] for a general review) and to reconstruct the distribution of dark matter.

CMB anisotropies. Any photon emitted during the decoupling and that is observed today has been redshifted due to the expansion of the universe but also because it propagates in an inhomogeneous spacetime. Besides, at emission, a denser region is hotter, the local velocity of the plasma is at the origin of a relative Doppler effect and the local gravitational potential induces an Einstein effect. The expression of the temperature fluctuation $\Theta = \delta T / T_{\mathrm{CMB}}$ observed here and today in a direction e) can be derived by solving the perturbed geodesic equation [87, 88] and leads to

$$\Theta(\boldsymbol{x}_0, \eta_0, \boldsymbol{e}) = \left[\frac{1}{4}\delta^{\mathrm{N}}{}_{\gamma} + \Phi - e^i \left(D_i V_{\mathrm{b}} + \bar{V}_{\mathrm{b}i} + \bar{\Phi}_i\right)\right]_{\mathrm{E}} , \bar{\eta}_{\mathrm{E}})$$
$$+ \int_{\mathrm{E}}^0 (\Phi' + \Psi') \left[\boldsymbol{x}(\eta), \eta\right] \mathrm{d}\eta - \int_{\mathrm{E}}^0 e^i \bar{\Phi}'_i [\boldsymbol{x}(\eta), \eta] \mathrm{d}\eta$$
$$- \int_{\mathrm{E}}^0 e^i e^j \bar{E}'_{ij} [\boldsymbol{x}(\eta), \eta] \mathrm{d}\eta , \tag{95}$$

where we integrate along the unperturbed geodesic that we chose to parameterise as

$$\boldsymbol{x} = \boldsymbol{x}_0 + \boldsymbol{e}(\eta_0 - \eta) , \tag{96}$$

so that the integral now runs over the conformal time η. This relation is known as the *Sachs-Wolfe equation*. It relates the observed temperature fluctuations to the cosmological perturbations. So $\Theta(\boldsymbol{x}_0, \eta_0, \boldsymbol{e})$ is a stochastic variable that can be characterized by its angular correlation function, C_ℓ (see, e.g., Refs [89] for early studies on the physics of the CMB and Refs. [90] for more recent formalisms).

We shall not derive its expression, but the computation runs through decomposing all the perturbation in Fourier modes and then as a transfer function and initial conditions, the observables in spherical harmonics. The average then acts on the initial conditions so that C_ℓ depends directly on the initial power spectrum and on the transfer function. It follows that its computation requires to solve the perturbed equation prior to decoupling that is for the photon-baryon-electron plasma coupled through Compton scattering.

The description of radiation however requires to go beyond a fluid description and to rely on a Boltzmann equation, that needs to be derived at linear order in a gauge invariant form (see, e.g., chapter 6 of Ref. [19] and references therein), but also beyond linear order [91]. This allows to determine, for any cosmological model, the angular correlation function of the temperature fluctuation, but also of the polarisation E- and B-modes, as well as the cross-correlation TE. Besides, the CMB photons are lens by the large scale structure of the universe, which offers a way to test the consistency of the whole model. This has recently been detected by the Planck experiment [92].

2.5.4. Beyond perturbation theory. As described above, the description of light propagation mostly relies on the cosmological perturbation theory [93]. At first order, it essentially introduces a dispersion of the distance-redshift relation with respect to the background FL prediction [94], which can be partially corrected if a lensing map is known. This problem of determining the effect of inhomogeneities on light propagation has also been tackled in a non perturbative way, e.g., relying on toy models. The most common examples are Swiss-cheese models [95].

However, when it comes to narrow beams, such as those involved in supernova observations, the approximation of describing the cosmic matter by a fluid does no longer hold. The applicability of the perturbation theory in this regime, in particular, has been questioned in Ref. [73].

This specific issue of how the *clumpiness* of the universe affects the interpretation of cosmological observables was first raised by Zel'dovich [96] and Feynman [97]. The basic underlying idea is that in a clumpy medium, light mostly propagates through vacuum, and should therefore experience an underdense universe. This stimulated a corpus of seminal articles [98] to question why the use of a FL spacetime is a good approximation to interpret the Hubble diagram (see Fig. 14. From a theoretical point it translates to the fact that in a FL universe, the Sachs equation is sourced only by the Ricci component (since the Weyl term vanishes) while in the true universe it is stress by the Weyl component (the Ricci term vanishing in vacuum).

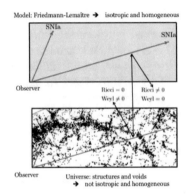
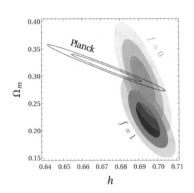

FIGURE 14. (left) The standard interpretation of SNe data assumes that light propagates in purely homogeneous and isotropic space (top). However, thin light beams are expected to probe the inhomogeneous nature of the actual universe (bottom) down to a scale where the continuous limit is no longer valid. (right) Comparison of the constraints obtained by *Planck* on (Ω_{m0}, H_0) [72] and from the analysis of the Hubble diagram constructed from the SNLS 3 catalog. The different contour plots correspond to different smoothness parameters f. For $f = 1$, the geometry used to fit the data is Friedmannian, as assumed in standard analysis. From Refs. [41, 42].

When applied to the interpretation of SN data, two effects have to be described: (1) the fact that in average the beam probes an underdense media and (2) the fact that each line of sight is different so that one also expect a larger dispersion. The first aspect was described by the Dyer-Roeder approximation but it does not tell anything about its dispersion, and a fortiori about its higher-order moments. The example of the analysis of Refs. [41, 42] shows that taking into account the effects of the small scale structure improves the agreement between SN and CMB observations regarding the measurement of Ω_m (see Fig. 14).

Recently, a new approach was proposed [99] and offers an analytical and a priori non-perturbative framework for determining the statistical impact of small-scale structures on light propagation. Its main idea is that, on very small scales, the matter density field (i.e., the source of lensing) can be treated a *white noise*, giving to lensing a diffusive behavior. The equations of geometric optics in curved spacetime then take the form of generalized Langevin equations, which come with the whole machinery of statistical physics. It then allows to write Fokker-Planck equation for the probability distribution function of the Jacobi matrix, which offers a very novel way to study the lensing in an inhomogeneous universe.

2.5.5. Summary. These examples show that relating observable to the perturbation variables and interpreting the effect of higher order correction consistently requires a deep theoretical insight. This steps cannot be avoided in order to determine how an observer in a given cosmological models sees his universe, which is what needs to be compared to the observations of our universe. Since all the predictions are from a statistical nature, one also needs to derive the variance of the observables.

This program is under control in a FL universe with linear perturbations, but many debates are open beyond linear order. One also needs to emphasize that as soon as the Copernican principle is not assumed, most of these computation, even at linear order, are difficult; see, e.g., Ref. [100] for the theory of linear perturbation in a Bianchi universe and Ref. [101] for the theory of weak-lensing.

2.6. Summary: the ΛCDM model

This description sets the basis of the standard cosmological model referred to as the ΛCDM model. In its minimal version it relies only on 6 cosmological parameters: 3 parameters (Ω_b, Ω_Λ, H_0) to describe the evolution of the background universe, assumed to be a spatially Euclidean FL spacetime (so that $\Omega_K = 0$, $\Omega_{cdm} = 1 - \Omega_b - \Omega_\Lambda$ and Ω_r being given by the mean temperature of the CMB and neutrinos), 2 parameters to describe the initial perturbations (amplitude and index of the power spectrum) and 1 parameter to describe the reionization.

Such a model is an excellent phenomenological model since it is in agreement with all observations. Deviations from it have been constrained, see, e.g., Ref. [72]. Dark matter strengthen the connection, that was already established by BBN, with particle physics.

This description draws however many questions. First, the initial power spectrum is purely ad-hoc and a further step in the construction of our model is to find its origin in the dynamics of the primordial universe. The dark energy and dark matter components requires to identify the degrees of freedom to which they are associated. This is a very active part of research at the crossroad between particle physics and cosmology. The lithium problem still exhibits an inconsistency that requires attention. Last but not least, the description of the efficiency of assuming that the universe is well-described by a FL spacetime requires also some attention. Indeed, one has replaced the spacetime metric tensor by an average metric of large scale –see Eq. (5)– and that the matter stress-energy tensor has been smoothed to solve the Einstein equation. Since the Einstein equations are not linear, the Einstein tensor of the smooth metric differs from the smooth Einstein tensor. One shall then question the use of the Friedmann equations and try to define the smoothing procedure. This remains an open question [102].

From a technical point of view, the use of perturbation theory to describe the large scale structure requires attention, in particular on its precision on small scales (see, e.g., Ref. [103]) where linear perturbation theory cannot be used. The question of the resummation of a higher perturbation theory is under question, as well a the way to properly treat the physics (inclusion of baryons, neutrinos,...). The common approach is to use this techniques together with N-body numerical simulations, which have their own limitations, such as the way to incorporate baryons and neutrinos, and also all relativistic effects; see, e.g., Ref. [104]. It is difficult to go below a 1% precision, which becomes a limitation for the precision cosmology program.

Besides, even the interpretation of some data, such as those related to thin beams, may require some caution and it is not clear that one can use the same metric to interpret all the observation, specially on small scale where the fluid limit does not hold [42].

3. The primordial universe

3.1. The inflationary paradigm

Inflation is defined as a primordial phase of accelerated of accelerated expansion, which has to last long enough for the standard problems of the hot big-bang model to be solved.

As explained above, it was initially proposed as a tentative solution to the problems of the hot big-bang model. In particular, it provides a simple explanation for the homogeneity and flatness of our universe.

Before the first models of inflation were proposed, precursor works appeared as early as 1965 by Erast Gliner [105], who postulated a phase of exponential expansion. In 1978, François Englert, Robert Brout and Egard Gunzig [106], in an attempt to resolve the primordial singularity problem and to introduce the particles and the entropy contained in the universe, proposed a 'fireball' hypothesis, whereby the universe itself would appear through a quantum effect in a state of negative pressure subject to a phase of exponential expansion. Alexei Starobinsky [107], in 1979, used quantum-gravity ideas to formulate the first semi-realistic rigorous model of an inflation era, although he did not aim to solve the cosmological problems. A simpler model, with transparent physical motivations, was then proposed by Alan Guth [32] in 1981. This model, now called 'old inflation', was the first to use inflation as a mean of solving cosmological problems. It was soon followed by Andrei Linde's 'new inflation' proposal [108].

From the Einstein equations, an accelerated expansion requires that the matter dominating the dynamics of the universe satisfies a relation between its energy density and pressure, $\rho + 3P < 0$. A solution to implement this condition is to use a scalar field φ. As long as it is slow-rolling, it satisfies $P \sim -\rho$ hence leading to acceleration, while if it is fast-rolling it satisfies $P \sim +\rho$. As long as this scalar field is a minimally coupled canonical field, the only freedom is the choice of its potential $V(\varphi)$.

Early models, known as *old inflation*, rely on a first-order phase-transition mechanism [109]; see Fig. 15. A scalar field is trapped in a local minimum of its potential, thus imposing a constant energy density, equivalent to the contribution of a cosmological constant. As long as the field remains in this configuration, the evolution of the universe is exponential and the universe can be described by a de Sitter spacetime. This configuration is metastable and the field can tunnel to its global minimum, $V(\varphi_f) = 0$, hence creating bubbles of true vacuum, which correspond to a non-inflationary universe. In models of *new inflation* [108], the scalar field exits its false vacuum by slowly rolling towards its true vacuum; see Fig. 15. These models can only work if the potential has a very flat plateau around $\varphi = 0$, which is artificial. In most of its versions, the inflaton cannot be in thermal equilibrium with other matter fields. The theory of cosmological phase transitions then does not apply and no satisfying realization of this model has been proposed so that it was progressively abandoned. The first models of inflation (see Ref. [110] for a review) were actually only incomplete modifications of the big-bang theory as they assumed that the universe was in a state of thermal equilibrium, homogeneous on large enough scales before the period of inflation. This problem was resolved by Linde with the proposition of chaotic inflation [111]. In this model, inflation can start from a Planckian density even if the universe is not in equilibrium. A new picture of the universe then appears. The homogeneity and isotropy of our observable universe would be only local properties, while the universe is very inhomogeneous on very large scales, with a fractal-like structure. Following the study of quantum effects near black-holes and in de Sitter spacetimes, quantum effects were investigated during inflation (see Ref. [112] for an historical prospect). Two approaches were followed. The first originating in the 60s [113], uses the static form of the de Sitter metric so that an observer at the origin would detect thermal radiation from $R = 1/H$ with a temperature $T = H/2\pi$, which corresponds to vacuum polarization of the de Sitter geometry. It is often used in the superstring community, in particular in the context of holography and the thermodynamics associated with horizons and the so-called "hot tin can" picture [114]. The second approach is based on quantization of a scalar field in a time-dependent background described by an (almost) de Sitter spacetime. This led to the formulation by Viatcheslav Mukhanov and Gennady Chibisov of the theory of cosmological perturbations during inflation [37, 115], which links the origin of the large scale structure of the universe to quantum fluctuations, quickly

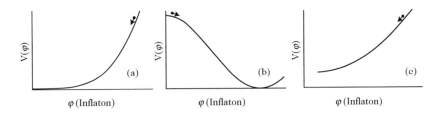

FIGURE 15. The model of old inflation (left) is based on a first
order phase transition from a local to a true minimum of the
potential. The false vacuum is metastable and the field can tunnel
to the true vacuum. In the models of new inflation (right) a scalar
field slowly relaxes towards its vacuum. From Ref. [19].

followed by a series of works [116, 117]. The slow-roll phase is essential in this
mechanism as it is during this period that the density fluctuations which lead to
the currently observed large scale structure, are generated.

At the end of inflation, all classical inhomogeneities have been exponentially
washed out, and one can consider them as non-existent at this stage. The universe
has become very flat so that curvature terms can be neglected. Moreover, all the
entropy has been diluted. If the inflaton potential has a minimum, the scalar field
will oscillate around this minimum right after the end of inflation. Due to the
Hubble expansion, these oscillations are damped and the scalar field decays into
a large number of particles. During this phase, the inflationary universe (of low
entropy and dominated by the coherent oscillations of the inflaton) becomes a hot
universe (of high entropy and dominated by radiation). This reheating phase [118]
connects inflation with the hot big-bang scenario and complete the picture. In
principle, knowing the couplings of the inflaton to the standard matter field, one
can determine the relative amount of all matter species and their distribution.
This step is however very challenging theoretically and requires heavy numerical
simulations.

3.2. Early motivations: the problems of the hot big-bang model

The different problems of the standard cosmological model are discussed in Chapters 3 and 5 of Ref. [19]. Let me start by the flatness and horizon problems in order
to show how inflation can solve it. To that purpose, let me rewrite the equations
of evolution in terms of the dimensionless parameter

$$\Omega_K = -\frac{K}{a^2 H^2}$$

as

$$\frac{\mathrm{d}\ln \Omega_K}{\mathrm{d}\ln a} = (1+3w)(1-\Omega_K)\Omega_K. \qquad (97)$$

It shows that Ω_K is a stable point of the dynamics which is not stable if $1+3w >$
0. It means that with standard pressureless matter and radiation, the curvature

term will tend to dominate the Friedmann equation at late time. But today, the spatial curvature of our universe is small $|\Omega_{K0} - 1| < 0.1$ (with the upper bound taken very generously). It implies that at the time of matter-radiation equality $|\Omega_K - 1| < 3 \times 10^{-5}$ and at Planck time $|\Omega_K - 1| < 10^{-60}$. The big-bang model does not give any explanation for such a small curvature at the beginning of the universe and this fine tuning is unnatural for an old universe like ours.

The cosmological principle, which imposes space to be homogeneous and isotropic, is at the heart of the FL solutions. By construction, these models cannot explain the origin of this homogeneity and isotropy. So it would be more satisfying to find a justification of this principle, at least on observable scales. A simple way to grasp the problem is to estimate the number of initial cells, with an initial characteristic Planck size length, present today in the observable universe. This number is typically of order

$$N \sim \left(\frac{1 + z_p}{\ell_P H_0}\right)^3 \sim 10^{87}.$$

The study of the CMB and of galaxies tends to show that their distribution is homogeneous on larger scales, so it is difficult to understand how initial conditions fixed on 10^{87} causally independent regions can appear so identical (at a 10^{-5} level!). This horizon problem is related to the state of thermodynamic equilibrium in which the universe is. The cosmological principle imposes a non-causal initial condition on spatial sections of the universe and in particular that the temperature of the thermal bath is the same at every point. The horizon problem is thus closely related to the cosmological principle and is therefore deeply rooted in the FL solutions.

3.3. Inflation

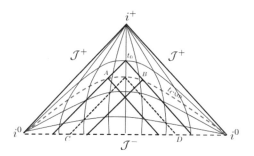

FIGURE 16. Penrose diagram of a universe with an intermediate stage of inflation (and no cosmological constant). It explains how such a period allow to solve the horizon problem. From Ref. [22].

3.3.1. The idea of inflation. Let us see how an phase of accelerated expansion can solve the flatness problem. Since the definition of Ω_K implies that $\Omega_K = -K/\dot{a}^2$, it is clear that it will (in absolute value) decrease to zero if $\dot{a}$ increases that is if $\ddot{a} > 0$, i.e., during a phase of accelerated expansion. As discussed above, this requires $\rho + 3P < 0$ and cannot be achieved with ordinary matter. This can also be seen on the Friedmann equations under the form (97). They clearly show that when $-1 < w < -\frac{1}{3}$, the fixed point $\Omega_K = 0$ is an attractor of the dynamics.

So, the flatness problem can be resolved if the attraction toward $\Omega_K = 0$ during the period of inflation is sufficient to compensate its subsequent drift away from 0 during the hot big-bang, i.e., if inflation has lasted sufficiently long. To quantify the duration of the inflationary period, we define the quantity

$$N \equiv \ln \left(\frac{a_f}{a_i} \right), \tag{98}$$

where a_i and a_f are the values of the scale factor at the beginning and at the end of inflation. This number measures the growth in the scale factor during the accelerating phase and is called the *number of "e-folds"*. To give an estimate of the required minimum number of e-folds of inflation, note that, if we assume H to be almost constant during inflation, then

$$\left| \frac{\Omega_K(t_f)}{\Omega_K(t_i)} \right| = \left(\frac{a_f}{a_i} \right)^{-2} = e^{-2N}.$$

In order to have $|\Omega_K(t_f)| \lesssim 10^{-60}$ and $\Omega_K(t_i) \sim \mathcal{O}(1)$, we thus need

$$N \gtrsim 70. \tag{99}$$

Similarly, in a accelerated universe, the comoving Hubble radius, $\mathcal{H}^{-1} = (aH)^{-1}$, decreases in time

$$\frac{\mathrm{d}}{\mathrm{d}t} (aH)^{-1} < 0. \tag{100}$$

Two points in causal contact at the beginning of inflation can thus be separated by a distance larger than the Hubble radius at the end of inflation. These points are indeed still causally connected, but can *seem* to be causally disconnected if the inflationary period is omitted. So, inflation allows for the entire *observable* universe to emerge out of the same causal region before the onset of inflation. The horizon problem can also be solved if $N \gtrsim 70$.

The Penrose diagram of a universe with a finite number of e-folds is depicted on Fig. 16.

3.3.2. Dynamics of single-field inflationary models.

Most models of inflation rely on the introduction of a dynamical scalar field φ, evolving in a potential, $V(\varphi)$ with an action given by

$$S = -\int \sqrt{-g} \left[\frac{1}{2} \partial_\mu \varphi \partial^\mu \varphi + V(\varphi) \right] \mathrm{d}^4 x, \tag{101}$$

so that its energy-momentum tensor takes the form

$$T_{\mu\nu} = \partial_\mu \varphi \partial_\nu \varphi - \left(\frac{1}{2} \partial_\alpha \varphi \partial^\alpha \varphi + V \right) g_{\mu\nu}. \tag{102}$$

It follows that the energy density and pressure of a homogenous scalar field are respectively given by

$$\rho_\varphi = \frac{\dot{\varphi}^2}{2} + V(\varphi), \quad P_\varphi = \frac{\dot{\varphi}^2}{2} - V(\varphi). \tag{103}$$

These expressions show that $\rho_\varphi + 3P_\varphi = 2(\dot{\varphi}^2 - V)$. Since the Friedmann equations imply that the expansion is accelerated as soon as $\dot{\varphi}^2 < V$. The expansion will be quasi-exponential if the scalar field is in slow-roll, i.e., if $\dot{\varphi}^2 \ll V$. This clearly explains why this is a natural way of implementing inflation.

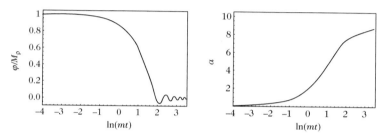

FIGURE 17. Evolution of the inflaton and scale factor during inflation with a potential (105). The scalar field is initially in a slow-roll regime and the expansion of the universe is accelerated. At the end of this regime, it starts oscillating at the minimum of its potential and it is equivalent to a pressureless fluid. From Ref. [19].

The Friedmann and Klein-Gordon equations take the form

$$H^2 = \frac{8\pi G}{3}\left(\frac{1}{2}\dot{\varphi}^2 + V\right) - \frac{K}{a^2}, \quad \frac{\ddot{a}}{a} = \frac{8\pi G}{3}\left(V - \dot{\varphi}^2\right),$$

$$\ddot{\varphi} + 3H\dot{\varphi} + V_{,\varphi} = 0. \tag{104}$$

Once the potential is chosen, the whole dynamics can be determined. As an example consider a free massive scalar

$$V(\varphi) = \frac{1}{2}m^2\varphi^2. \tag{105}$$

The Klein-Gordon equation reduces to that of a damped harmonic oscillator. If φ is initially large, then the Friedmann equation implies that H is also very large. The friction term becomes important and dominates the dynamics so that the field must be in the slow-roll regime. The evolution equations then reduce to

$$3H\dot{\varphi} + m^2\varphi = 0, \quad H^2 = \frac{4\pi}{3}\left(\frac{m}{M_p}\right)^2\varphi^2,$$

that give

$$\varphi(t) = \varphi_{\rm i} - \frac{mM_p}{\sqrt{12\pi}}t, \quad a(t) = a_{\rm i}\exp\left\{\frac{2\pi}{M_p^2}\left[\varphi_{\rm i}^2 - \varphi^2(t)\right]\right\}, \tag{106}$$

where φ_i and a_i are the values of the field and the scale factor at $t_i = 0$ and M_p is the Planck mass. This can be compared to the numerical integration depicted in Fig. 17, in which we see the subsequent (non-slow-rolling) phase with the damped oscillations at the bottom of the potential and the fact that $\ddot{a}$ changes sign to become negative. Figure 18 provides a phase space analysis of the dynamics, showing that the slow-roll trajectories are attractors of the dynamics. It follows that if inflation lasts long enough, the initial conditions on $(\varphi, \dot{\varphi})$ become irrelevant.

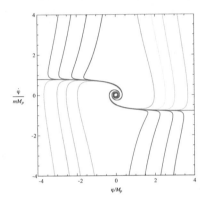

FIGURE 18. Phase portrait in the $(\varphi, \dot{\varphi})$-plane of the dynamics of a scalar field with potential $V = m^2 \varphi^2/2$, assuming $m = 10^{-6} M_p$. This illustrates the mechanism of attraction toward the slow-roll solution. From Ref. [19].

3.3.3. Slow-roll formalism. As seen of the example of a massive scalar field, inflation occurs while the scalar field is slow-rolling. This has led to the development of a perturbative formalism to describe the dynamics. If the field is slow-rolling then H is almost constant and it is convenient to define

$$\varepsilon = -\frac{\dot{H}}{H^2}, \quad \delta = \varepsilon - \frac{\dot{\varepsilon}}{2H\varepsilon}, \quad \xi = \frac{\dot{\varepsilon} - \dot{\delta}}{H}. \tag{107}$$

These definitions depend only on the spacetime geometry. In the case of a single scalar field, they can be rewritten as

$$\varepsilon = \frac{3}{2}\dot{\varphi}^2 \left[\frac{1}{2}\dot{\varphi}^2 + V(\varphi)\right]^{-1}, \quad \delta = -\frac{\ddot{\varphi}}{H\dot{\varphi}}. \tag{108}$$

ε can be used to rewrite the Friedmann equations as

$$H^2 \left(1 - \frac{1}{3}\varepsilon\right) = \frac{8\pi G}{3}V, \quad \frac{\ddot{a}}{a} = H^2 (1 - \varepsilon), \tag{109}$$

and the effective equation of state of the inflaton as $w_\varphi = -1 + \frac{2}{3}\varepsilon$ so that he condition for inflation reduces

$$\ddot{a} > 0 \iff w < -1/3 \iff \varepsilon < 1. \tag{110}$$

The number of e-folds between a time t where the value of the inflaton is φ and the end of inflation ($t = t_{\mathrm{f}}$ and $\varphi = \varphi_{\mathrm{f}}$, of the inflationary phase can be expressed as

$$N(t, t_{\mathrm{f}}) = \int_{t}^{t_{\mathrm{f}}} H \, \mathrm{d}t \,, \tag{111}$$

since, after integration, $a(t) = a_{\mathrm{f}} \, e^{-N}$. $N(t_{\mathrm{i}}, t_{\mathrm{f}})$ corresponds to the duration of the inflationary phase, as defined earlier. N can be expressed as

$$N(\varphi, \varphi_{\mathrm{f}}) = \int_{\varphi}^{\varphi_{\mathrm{f}}} \frac{H}{\dot{\varphi}} \, \mathrm{d}\varphi = -\sqrt{4\pi G} \int_{\varphi}^{\varphi_{\mathrm{f}}} \frac{\mathrm{d}\varphi}{\sqrt{\varepsilon}} \,. \tag{112}$$

As long as the slow-roll parameters are small, H is almost constant. One can then develop the equations for the evolution of the background (a, H, φ) in terms of this small parameters. This allows to derive the observational predictions in term of these parameters and, interestingly they can be related to derivative of the inflationary potential [19, 119] as

$$\varepsilon = \frac{1}{16\pi G} \left(\frac{V_{,\varphi}}{V} \right)^{2} \,, \quad \delta = \frac{1}{8\pi G} \left(\frac{V_{,\varphi\varphi}}{V} \right) - \varepsilon.$$

3.3.4. Chaotic inflation. In order to illustrate this formalism, let us come back to the massive scalar field. It is clear that $\varepsilon = \frac{M_{p}^{2}}{4\pi\varphi^{2}}$ and $\delta = 0$, so that the slow-roll regime lasts until φ reaches $\varphi_{\mathrm{f}} = M_{p}/\sqrt{4\pi}$. We infer that the total number of e-folds is

$$N(\varphi_{\mathrm{i}}) = 2\pi \left(\frac{\varphi_{\mathrm{i}}}{M_{p}} \right)^{2} - \frac{1}{2} \,. \tag{113}$$

In order to have $N \gtrsim 70$, we need $\varphi_{\mathrm{i}} \gtrsim 3M_{p}$. If φ takes the largest possible value compatible with the classical description adopted here, i.e., $V(\varphi_{\mathrm{i}}) \lesssim M_{p}^{4}$, we find that $\varphi_{\mathrm{i}} \sim M_{p}^{2}/m$. In this case, the maximal accessible number of e-folds would be $N_{\mathrm{max}} \sim 2\pi M_{p}^{2}/m^{2}$. As can been deduced from the observations of the CMB impose that $m \sim 10^{-6} M_{p}$, so that $N_{\mathrm{max}} \sim 10^{13}$. The maximal number of e-folds is thus very large compared to the minimum required for solving the cosmological problems.

Consequently, if the universe is initially composed of regions where the values of the scalar field are randomly distributed then the domains where the initial value of φ is too small never inflate or only for a small number of e-folds. The main contribution to the total physical volume of the universe at the end of inflation comes from regions that have inflated for a long time and that had an initially large value of φ. These domains produce extremely flat and homogeneous zones at the end of inflation with a very large size compared to that of the observable universe.

This is the idea of *chaotic inflation* proposed in Ref. [111] and which predicts, if we are typical observers, that we shall not be surprised to observe an extremely

Euclidean, homogeneous and isotropic universe. This conclusion has to be contrasted with the early version of the big-bang model and to the discussion on its problems.

3.3.5. End of the inflationary phases. Inflation ends when $\max(\varepsilon, \delta) \sim 1$. At the end of inflation, all classical inhomogeneities have been exponentially washed out, and one can consider them as non-existent at this stage (still the fact that we neglect them from the start of the analysis has motivated many works to carefully understand the homogenization and isotropization of the universe, both at the background and perturbative levels; see, e.g., Ref. [120] for a discussion of the curvature and Ref. [121] for a discussion on the isotropization and their effects on the maximum number of e-folds) so that one can set $K = 0$ during all the late stages of the primordial phase.

During inflation, all the energy is concentrated in the inflaton. Shortly after the end of inflation, the universe is cold and "frozen" in a state of low entropy where the field oscillates around the minimum of its potential. The coherent oscillations of the inflaton can be considered as a collection of independent scalar particles. If they couple to other particles, the inflaton can decay perturbatively to produce light particles. The interaction of the inflaton should therefore give rise to an effective decay rate, Γ_φ, and reheating would only occur after the expansion rate had decreased to a value $H \sim \Gamma_\varphi$. This also implies that during the first $\mathcal{O}(m/\Gamma_\varphi)$ oscillations of the inflaton, nothing happens.

To illustrate this consider an inflaton with potential $V \propto \varphi^n$. During the oscillatory phase, $H < m$ and the inflaton undergoes several oscillations during a time H^{-1}. It is thus reasonable to use the mean value of the pressure and density over several oscillations. It follows that $\langle \dot\varphi^2/2 \rangle = (n/2)\langle V(\varphi) \rangle$, so that $\langle P_\varphi \rangle = (n-2)/(n+2)\langle \rho_\varphi \rangle$. The scalar field thus behaves as a dust fluid for $n = 2$ and as a radiation fluid for $n = 4$.

In order for the inflaton to decay, it should be coupled to other fields. Including quantum corrections, the Klein-Gordon equation becomes

$$\ddot\varphi + 3H\dot\varphi + \left[m^2 + \Pi(m)\right]\varphi = 0, \tag{114}$$

where $\Pi(m)$ is the polarisation operator of the inflaton. The real part of $\Pi(m)$ corresponds to the mass correction. Π has an imaginary part $\mathrm{Im}[\Pi(m)] = m\Gamma_\varphi$. Since m is much larger than both Γ_φ and H at the end of inflation, we can solve the Klein-Gordon equation by assuming that both these quantities are constant during an oscillation. It follows that

$$\varphi = \Phi(t)\sin mt, \quad \Phi = \varphi_0 \exp\left[-\frac{1}{2}\int(3H + \Gamma_\varphi)\mathrm{d}t\right]. \tag{115}$$

As long as $3H > \Gamma_\varphi$, the decrease in the inflaton energy caused by the expansion (Hubble friction) dominates over particle decay. Thus,

$$\Phi = \varphi_{\mathrm{f}}\frac{t_{\mathrm{f}}}{t} = \frac{M_p}{m}\frac{1}{\sqrt{3\pi}t} \tag{116}$$

where we have used $\varphi_f = M_p/\sqrt{4\pi}$, $t_f = 2/3H_f$ and $H_f^2 = (4\pi/3)(m/M_p)^2\varphi_f^2$. Reheating occurs in the regime $\Gamma_\varphi \gtrsim 3H$ and

$$\Phi = \frac{M_p}{m}\frac{1}{\sqrt{3\pi t}}e^{-\Gamma_\varphi t/2}. \tag{117}$$

We define the time of reheating, t_{reh}, by $\Gamma_\varphi = 3H$ so that the energy density at that time is $\rho_{\text{reh}} = \frac{\Gamma_\varphi^2 M_p^2}{24\pi}$. If this energy is rapidly converted into radiation, its temperature is $\rho_{\text{reh}} = \frac{\pi^2}{30}g_* T_{\text{reh}}^4$. This defines the reheating temperature ias

$$T_{\text{reh}} = \left(\frac{5}{4\pi^3 g_*}\right)^{1/4}\sqrt{\Gamma_\varphi M_p} \simeq 0.14\left(\frac{100}{g_*}\right)^{1/4}\sqrt{\Gamma_\varphi M_p} \ll 10^{15}\,\text{GeV}, \tag{118}$$

where the upper bound was obtained from the constraint $\Gamma_\varphi \ll m \sim 10^{-6}M_p$, assuming $g_* \gtrsim 100$ for the effective number of relativistic particles.

This description of perturbative reheating is simple and intuitive in many aspects. However, the decay of the inflaton can start much earlier in a phase of *preheating* (parametric reheating) where particles are produced by parametric resonance. The preheating process can be decomposed into three stages: (1) non perturbative production of particles, (2) perturbative stage and (3) thermalization of the produced particles. We refer to Ref. [118] for details on this stage that connects inflation to the standard hot big-bang model.

3.4. A scenario for the origin of the large scale structure

The previous section has described a homogeneous scalar field. As any matter field, φ has quantum fluctuations and cannot be considered as strictly homogeneous. This drives us to study the effects of these perturbations. Any fluctuation of the scalar field will generate metric perturbations since they are coupled by the Einstein field equations. We should thus study the coupled inflaton-gravity system to understand.

In terms of the variables to consider, we start from 10 perturbations for the metric and 1 for the scalar field, to which we have to subtract 4 gauge freedoms and 4 constraint equations (2 scalars and 2 vectors). We thus expect to identify 3 independent degrees of freedom to describe the full dynamics: one scalar mode and a tensor mode (counting for 2 degrees of freedom, one per polarization).

The following presentation relies on the extended description of Ref. [19] (chapter 8).

3.4.1. Perturbation theory during inflation.
Starting from the perturbation of the scalar field as $\varphi = \varphi(t) + \delta\varphi(\boldsymbol{x}, t)$, its stress-energy tensor takes the form

$$\delta T_{\mu\nu} = 2\partial_{(\nu}\varphi\partial_{\mu)}\delta\varphi - \left(\frac{1}{2}g^{\alpha\beta}\partial_\alpha\varphi\partial_\beta\varphi + V\right)\delta g_{\mu\nu}$$

$$- g_{\mu\nu}\left(\frac{1}{2}\delta g^{\alpha\beta}\partial_\alpha\varphi\partial_\beta\varphi + g^{\alpha\beta}\partial_\alpha\delta\varphi\partial_\beta\varphi + V'\delta\varphi\right). \tag{119}$$

One can then use the standard theory of gauge invariant cosmological perturbation described above. To that purpose, one needs introduce the two following gauge invariant variables for the scalar field

$$\chi = \delta\varphi + \varphi'(B - E'), \quad \text{or} \quad Q = \delta\varphi - \varphi'\frac{C}{\mathcal{H}}, \tag{120}$$

They are related by $Q = \chi + \varphi'\Psi/\mathcal{H}$. and χ is the perturbation of the scalar field in Newtonian gauge while Q, often called the Mukhanov-Sasaki variable is the one in flat slicing gauge.

Forgetting about the technicalities of the derivation, one ends up with the following equations of evolution. We were left with $7 = 2+2+3$ degrees of freedom respectively fo the T, V and S modes. For simplicity we assume $K = 0$.

- *Vector modes.* Since a scalar field does not contain any vector sources, there are no vector equations associated with the Klein-Gordon equation and there is only one Einstein equation for these modes and a constraint equation,

$$\Delta\bar{\Phi}_i = 0.$$

We can conclude that $\bar{\Phi}_i = 0$ independently of any model of inflation, that the vector modes are completely absent at the end of inflation. The 2 constraints equation kill the vector degrees of freedom.

- *Tensor modes.* There is only one tensor equation which can be obtained from the Einstein equations

$$\bar{E}_{ij}'' + 2\mathcal{H}\bar{E}_{ij}' - \Delta\bar{E}_{ij} = 0. \tag{121}$$

It describes the evolution of the two polarisations of a gravity wave as a damped harmonic oscillator. Shifting to Fourier space and decomposing the gravity waves on a polarization tensor as $\bar{E}_{ij}(\boldsymbol{k}, , \eta) = \sum_\lambda \bar{E}_\lambda(\boldsymbol{k}, \eta)\varepsilon_{ij}^\lambda(\boldsymbol{k})$ and defining

$$\mu_\lambda(\boldsymbol{k}, \eta) = \sqrt{\frac{M_p^2}{8\pi}}a(\eta)\bar{E}_\lambda(\boldsymbol{k}, \eta)$$

the equation of evolution takes the form

$$\mu_\lambda'' + \left(k^2 - \frac{a''}{a}\right)\mu_\lambda = 0 \tag{122}$$

for each of the two polarizations.

- *Scalar modes.* The Einstein equations provide 2 constraint equations and an equation of evolution. This implies that we shall be able to derive a master equation for the only propagating degree of freedom. This can be achieved by defining [115, 19]

$$v(\boldsymbol{k}, \eta) = a(\eta)Q(\boldsymbol{k}, \eta), \quad z(\eta) = \frac{a\varphi'}{\mathcal{H}}$$

that can be shown to satisfy

$$v'' + \left(k^2 - \frac{z''}{z}\right)v = 0. \tag{123}$$

Bote that v is a composite variable of the field and metric perturbations. As expected from general relativity, it is not possible to separate them.

The set of equations (122-123) for the tensor and scalar modes are second order differential equations in Fourier space. In order for them to be predictive, one needs to fix their initial conditions. It is striking that these equations look like Schrödinger equations with a time dependent mass.

3.4.2. Setting the initial conditions. Let us first compare the evolution of a mode with comoving wave-number k in the standard big-bang model and during inflation. Any equation of evolution can be shown to have two regime whether the mode is super-Hubble $(k < \mathcal{H})$ or sub-Hubble $(k > \mathcal{H})$; see Fig. 19.

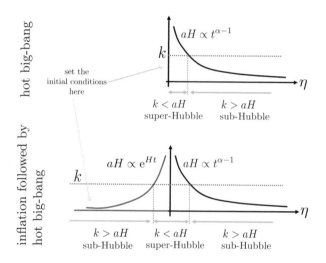

FIGURE 19. Evolution of the comoving Hubble radius and of a comoving mode of wavenumber k with conformal time η in the standard hot big-bang model (top) and with an inflationary phase (bottom). Without inflation, the mode is always super-Hubble in the past and becomes sub-Hubble as the universe expands. This implies that initial conditions have to be set on super-Hubble scales. The existence of super-Hubble correlation are thus thought to be acausal. With an inflationary phase, the mode was sub-Hubble deep in the inflationary era, which means that initial conditions have to be set on sub-Hubble scales.

In the post-inflationary era, the expansion is decelerating so that $\mathcal{H}$ is a decreasing function of η while k remains constant (since it is comoving). It follows that the super-Hubble modes become sub-Hubble while the universe expands. All observable modes are sub-Hubble today and one needs to set initial conditions in the early universe while they were super-Hubble. Actually, asymptotically toward the big-bang all modes where super-Hubble. A difficulty arises from the observation of the CMB and large scale structure that show that there shall exist super-Hubble correlations. It seems unnatural to set correlated initial conditions on super-Hubble scales since it would appear as acausal. Besides, there was no natural procedure to fix the initial conditions so that they were completely free and reconstructed to get an agreement with the observations.

During inflation, the picture is different because the expansion is accelerated, which means that $\mathcal{H}$ is increasing with η while k remains constant. It follows that any super-Hubble modes was sub-Hubble deep in the inflationary era. If inflation lasts long enough then all the observable modes can be sub-Hubble where v and μ_λ behave as harmonic oscillator. We shall thus find a mechanism to set initial conditions in this regime since we are interested only in modes that are observable today, at least as far as confronting inflation to observation is concerned.

The idea of Ref. [37] was to take seriously the fact that Q, and thus v, enjoys quantum fluctuations. They demonstrated that when expanded at second order in perturbation the Einstein Hilbert + scalar field action reduces (see Refs. [115, 121, 13]) to

$$\delta^{(2)} S = \frac{1}{2} \int \mathrm{d}\eta \mathrm{d}^3 \boldsymbol{x} \left[(v')^2 - \delta^{ij} \partial_i v \partial_j v + \frac{z''}{z} v^2 \right] \equiv \int \mathcal{L} \mathrm{d}^4 x , \qquad (124)$$

up to terms involving a total derivative and that do not contribute to the equations of motion. The variation of this action indeed gives the equation (123) but it tells us more on the structure of the theory: we recognize the action of a canonical scalar field with a time-dependent mass, $m^2 = -z''/z$ in Minkowski spacetime. It was thus proposed that the variable to quantize is v and we shall quantize it as we quantize any canonical scalar field evolving in an time-dependent exterior field [122, 123], where here the time-dependence would find its origin in the spacetime dynamics [124].

The procedure then goes as follows

1. v is promoted to the status of quantum operator in second quantization in Heisenberg representation

$$\hat{v}(\boldsymbol{k}, \eta) = \int \frac{\mathrm{d}^3 \boldsymbol{k}}{(2\pi)^{3/2}} \left[v_k(\eta) \mathrm{e}^{i\boldsymbol{k}\cdot\boldsymbol{x}} \hat{a}_{\boldsymbol{k}} + v_k^*(\eta) \mathrm{e}^{-i\boldsymbol{k}\cdot\boldsymbol{x}} \hat{a}_{\boldsymbol{k}}^\dagger \right] , \qquad (125)$$

where $\hat{a}_{\boldsymbol{k}}$ and $\hat{a}_{\boldsymbol{k}}^\dagger$ are creation and annihilation operators.
2. One introduces he conjugate momentum of v,

$$\pi = \frac{\delta \mathcal{L}}{\delta v'} = v' , \qquad (126)$$

which is also promoted to the status of operator, $\hat{\pi}$.
3. One can then get the Hamiltonian

$$H = \int (v'\pi - \mathcal{L}) \, \mathrm{d}^4 x = \frac{1}{2} \int \left(\pi^2 + \delta^{ij} \partial_i v \partial_j v - \frac{z''}{z} v^2 \right) \mathrm{d}^4 x . \qquad (127)$$

The equation of evolution for $\hat{v}$ is indeed Eq. (123) which is equivalent to the Heisenberg equations $\hat{v}' = i\left[\hat{H}, \hat{v}\right]$ and $\hat{\pi}' = i\left[\hat{H}, \hat{\pi}\right]$.

4. The operators $\hat{v}$ and $\hat{\pi}$ have to satisfy canonical commutation relations on constant time hypersurfaces

$$[\hat{v}(\boldsymbol{x},\eta), \hat{v}(\boldsymbol{y},\eta)] = [\hat{\pi}(\boldsymbol{x},\eta), \hat{\pi}(\boldsymbol{y},\eta)] = 0, \quad [\hat{v}(\boldsymbol{x},\eta), \hat{\pi}(\boldsymbol{y},\eta)] = i\delta(\boldsymbol{x} - \boldsymbol{y}). \quad (128)$$

5. As for quantization in Minkowski spacetime, the creation and annihilation operators appearing in the decomposition (125) satisfy the standard commutation rules

$$[\hat{a}_{\boldsymbol{k}}, \hat{a}_{\boldsymbol{p}}] = [\hat{a}_{\boldsymbol{k}}^\dagger, \hat{a}_{\boldsymbol{p}}^\dagger] = 0, \quad [\hat{a}_{\boldsymbol{k}}, \hat{a}_{\boldsymbol{p}}^\dagger] = \delta(\boldsymbol{k} - \boldsymbol{p}). \quad (129)$$

They are consistent with the commutation rules (128) only if v_k is normalized according to

$$W(k) \equiv v_k v'^*_k - v_k^* v'_k = i, \quad (130)$$

This determines the normalisation of their Wronskian, W.

6. The vacuum state $|0\rangle$ is then defined by the condition that it is annihilated by all the operators $\hat{a}_{\boldsymbol{k}}$, $\forall\, \boldsymbol{k}$, $\hat{a}_{\boldsymbol{k}}\,|0\rangle = 0$. The sub-Hubble modes, i.e., the high-frequency modes compared to the expansion of the universe, must behave as in a flat spacetime. Thus, one decides to pick up the solution that corresponds adiabatically to the usual Minkowski vacuum

$$v_k(\eta) \to \frac{1}{\sqrt{2k}} e^{-ik\eta}, \quad k\eta \to -\infty. \quad (131)$$

This choice is called the *Bunch-Davies vacuum*. Let us note that in an expanding spacetime, the notion of time is fixed by the background evolution which provides a preferred direction. The notion of positive and negative frequency is not time invariant, which implies that during the evolution positive frequencies will be generated.

Note that this quantization procedure amounts to treating gravity in a quantum way at linear order since the gravitational potential is promoted to the status of operator.

It can be seen that it completely fixes the initial conditions, i.e., the two free functions of integration that appear when solving Eq. (123). This can be seen on the simple example of a de Sitter phase ($H = $ constant, $a(\eta) = -1/H\eta$ with a test scalar field. Then $z''/z = 2/\eta$ and the equation (123) simplifies to $v_k'' + \left(k^2 - \frac{2}{\eta^2}\right)v_k = 0$ for which the solutions are given by

$$v_k(\eta) = A(k)e^{-ik\eta}\left(1 + \frac{1}{ik\eta}\right) + B(k)e^{ik\eta}\left(1 - \frac{1}{ik\eta}\right). \quad (132)$$

The condition (131) sets $B(k) = 0$ and fixes the solution to be (for $Q = v/a = H\eta v$)

$$Q_k = \frac{H\eta}{\sqrt{2k}}\left(1 + \frac{1}{ik\eta}\right)e^{-ik\eta}. \quad (133)$$

Notice that on sub-Hubble $Q_k \propto 1/\sqrt{k}$ as imposed by quantum mechanics but when the mode becomes super-Hubble ($k\eta \ll 1$) it shifts to $Q_k \propto 1/\sqrt{k^3}$, which corresponds to a scale invariant power spectrum. Hence the scale invariance on super-Hubble scales is inherited from the small scale properties fixed by quantum mechanics and no tuning is at work.

On super-Hubble scales, the field Q acquires a constant amplitude $|Q_k|$ ($k\eta \ll 1$) $= \frac{H}{\sqrt{2k^3}}$. In this limit

$$\hat{Q} \to \int \frac{\mathrm{d}^3 \boldsymbol{k}}{(2\pi)^{3/2}} \, \hat{\chi}_{\boldsymbol{k}} \, e^{i\boldsymbol{k}\cdot\boldsymbol{x}} = \int \frac{\mathrm{d}^3 \boldsymbol{k}}{(2\pi)^{3/2}} \frac{H}{\sqrt{2k^3}} \left(\hat{a}_{\boldsymbol{k}} + \hat{a}^\dagger_{-\boldsymbol{k}} \right) e^{i\boldsymbol{k}.\boldsymbol{x}}$$

All the modes being proportional to $(\hat{a}_{\boldsymbol{k}} + \hat{a}^\dagger_{-\boldsymbol{k}})$, $\hat{Q}$ has the same statistical properties as a Gaussian classical stochastic field and its power spectrum defined by

$$\langle Q_{\boldsymbol{k}} Q^*_{\boldsymbol{k}'} \rangle = P_Q(k) \, \delta^{(3)}(\boldsymbol{k} - \boldsymbol{k}') \,, \tag{134}$$

from which one easily deduces that

$$P_Q(k) = \frac{2\pi^2}{k^3} \mathcal{P}_Q(k) = |Q_k|^2 = \frac{|v_k|^2}{a^2} \,. \tag{135}$$

It is thus a scale invariant power spectrum since $\mathcal{P}_Q(k) = \left(\frac{H}{2\pi}\right)^2$.

3.5. Generic predictions and status

The analysis described in the previous section applies in any model of inflation and the conclusions can be computed analytically in the slow-roll regime. The departure from a pure de Sitter phase reflects itself in a spectral index in Eq. (135). They also apply identically for the gravity waves, the only difference being that v and μ_λ have different "mass terms" and that they are related to Q and $\bar{E}_\lambda$ respectively by z and a. It implies that they will have different spectral indices and amplitudes.

in conclusion, single field inflationary models have robust predictions that are independent of their specific implementation.

- The observable universe must be homogeneous and isotropic. Inflation erases any classical inhomogeneities. The observable universe can thus be described by a FL spacetime.
- The universe must be spatially Euclidean. During inflation, the curvature of the universe is exponentially suppressed. Inflation therefore predicts that $\Omega_K = 0$ up to the amplitude of the super-Hubble density perturbations, i.e., up to about 10^{-5}.
- Scalar perturbations are generated. Inflation finds the origin of the density perturbations in the quantum fluctuations of the inflaton which are amplified and redshifted to macroscopic scales. All the modes corresponding to observable scales today are super-Hubble at the end of inflation. Inflation predicts that these perturbations are adiabatic, have a Gaussian statistics and have an almost scale-invariant power spectrum. The spectral index can vary slightly with wavelength. These perturbations are coherent, which is translated into

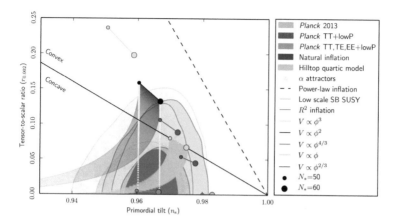

FIGURE 20. Constraints on inflationary models from Planck: marginalized joint 68% and 95% confidence level regions for the spectral index n_s and scalar-to-tensor ratio $r_{0.002}$ from Planck in combination with other data sets, compared to the theoretical predictions of selected inflationary models. From Ref. [125].

a structure of acoustic peaks in the angular power spectrum of the CMB's temperature anisotropies.

- There are no vector perturbations.
- Gravitational waves are generated. In the same way as for scalar modes, the gravitational waves have a quantum origin and are produced through parametric amplification. They also have Gaussian statistics and an almost scale invariant power spectrum.
- There is a consistency relation between their spectral index and the relative amplitude of the scalar and tensor modes.
- The quantum fluctuations of any light field $(M < H)$ are amplified. This field develops super-Hubble fluctuations of amplitude $H/2\pi$.

Note that this can be viewed as a no-hair theorem for inflation since any classical perturbations prior to inflation are erased while quantum fluctuations survive.

More generally, inflationary models radically change our vision of cosmology in at least three respects. These predictions are in full agreement with observations, and CMB in particular (see Fig. 20).

- Predictions are by essence probabilistic. One can only predict spectra, correlations so that only the statistical properties of the galaxy distribution can be inferred, not their exact position.
- Particles are produced by the preheating mechanism. The inflaton decay allows us to explain the production of particles at the end of inflation. It may affect the standard predictions [126].

- Inflation is eternal. Placing inflationary models in the context of chaotic initial conditions, we obtain a very different picture of the universe. The latter should be in eternal inflation and gives rise to island universes.

3.6. Extrapolation

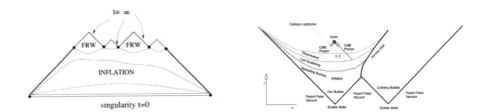

FIGURE 21. Penrose diagrams for a universe with eternal inflation (left) and with nucleation of bubble universes (right). Respectively from Ref. [112] and Ref. [127].

3.6.1. Eternal inflation. To finish, let us describe the process of eternal inflation that plays a central role in the discussion of the multiverse. The discovery of the self-reproduction process is a major development for inflation and cosmology. This mechanism was known for the models of old and new inflation and then extended to chaotic inflation [128]. In inflationary models with large values of the inflaton, quantum fluctuations can locally increase the value of the inflaton. The expansion of these regions is then faster and their own quantum fluctuations generate new inflationary domains. This process naturally leads to a self-reproducing universe where there are always inflating regions.

We just sketch a heuristic description here [129]. Regions separated by distances greater than H^{-1} can be considered to be evolving independently. So any region of size H^{-1} will be considered as an independent universe decoupled from other regions. Consider such a region of size H^{-1} for which the scalar field is homogeneous enough and has a value $\varphi \gg M_p$ and consider a massive field. In a time interval $\Delta t \sim H^{-1}$, the field classically decreases by $\Delta \varphi \sim \dot{\varphi} \Delta t \sim \dot{\varphi}/H$. The Klein-Gordon equation in the slow-roll regime then implies that $\Delta \varphi \sim -M_p^2/4\pi\varphi$. This value should be compared with the typical amplitude of quantum fluctuations, $|\delta\varphi| \sim H/2\pi \sim m\varphi/\sqrt{3\pi}M_p$ (see next section).

Classical and quantum fluctuations have the same amplitude for

$$|\Delta\varphi| \sim |\delta\varphi| \iff \varphi \sim \varphi_* \equiv \frac{M_p}{2}\sqrt{\frac{M_p}{m}}. \qquad (136)$$

We can thus distinguish between three phases in the evolution of the inflaton: a phase during which the quantum fluctuations are of the same order (or larger) as the classical field variation, a phase in which the field is in classical slow-roll towards its minimum and a phase when the field oscillates around its minimum.

Note that $V(\varphi_*) = (m/M_p)M_p^4/8 \lesssim 10^{-6}M_p^4$ so that the first regime can be reached even at energies small compared to M_p.

When $\varphi \gg \varphi_*$, $\delta\varphi \gg \Delta\varphi$. The characteristic length of the fluctuations $\delta\varphi$ generated during the time Δt is of the order of H^{-1} so that the initial volume is divided into $(\exp H\Delta t)^3 \sim 20$ independent volumes of radius H^{-1}. Statistically the value of the field in half of these regions is $\varphi + \Delta\varphi - \delta\varphi$ and $\varphi + \Delta\varphi + \delta\varphi$ in the other half. So the physical volume of the regions where the field has a value greater than φ_* is ten times larger

$$V_{t+\Delta t}(\varphi > \varphi_*) \sim \frac{1}{2}\left(\exp H\Delta t\right)^3 V_t(\varphi > \varphi_*) \sim 10 V_t(\varphi > \varphi_*). \qquad (137)$$

The physical volume where the space is inflating therefore grows exponentially in time. The zones where the field becomes lower than φ_* enter a slow-roll phase. So they become inflationary universes, decoupled from the rest of the universe, with a slow-roll phase, a reheating phase and a hot big-bang. These zones are *island universes* (or pocket universes) and our observable universe would be only a tiny part of such an island universe.

This scenario has important consequences for cosmology. In chaotic inflation, the universe has a very inhomogeneous structure on scales larger than H^{-1} with regions undergoing eternal inflation continuously giving rise to new zones themselves undergoing inflation. On large scales, the universe has a fractal structure with continuous production of island universes. Each one of these island universes then undergoes a phase of "classical" inflation, with a large number of e-folds and is thus composed of many regions of the size of our observable universe. This simplistic model assumes only one scalar field and a potential with a unique minimum. Realistic models in high energy physics on the other hand involve many scalar fields. The potential of these fields can be very complex and have many flat directions and minima. So the same theory can have different vacua which correspond to different schemes of symmetry breaking. Each of these vacua can lead, at low energy, to physically different laws. Due to the exploration of this *landscape* by quantum fluctuations, the universe would find itself divided into many regions with different low-energy physical laws (for instance different values of the fundamental constants). If this vision of the primordial universe is correct, physics alone cannot provide a complete explanation for all the properties of our observable universe since the same physical theory can generate vast regions with very different low energy properties. So our observable universe would have the properties it has not because the other possibilities are impossible or improbable, but simply because a universe with such properties allows for a life similar to ours to appear.

Eternal inflation thus offers a framework to apply the anthropic principle since the self-reproduction mechanism makes it possible to generate universes with different properties and to explore all possible vacua of a theory. This approach is used more and more to address the question of the value of the fundamental constants and to address of the cosmological constant problem. This framework also allows us to address questions beyond the origin of the properties of our universe thus defining the limitation of what we will be capable of explaining.

The Penrose diagram in eternal inflation is depicted on Fig. 21 and compared to another scenario based on bubble nucleation. In the case the $\mathcal{I}^+$ hypersurface has a fractal structure. As observers, we live in one of the late time asymptotic FL pockets.

4. Conclusions

Our current cosmological model can be considered as a simple and efficient model to explain the dynamics of the universe and of the structures it contains. It relies on 4 main hypothesis on the laws of nature and the symmetries of the geometry of the cosmological solutions. The Copernican principle thus plays a central role in this construction. The model is structured in different layers, that have historically been developed one after the others: (1) cosmological solution of general relativity, (2) the hot big-bang model, (3) the description of the large scale structure and (4) a scenario of the primordial universe and of the origin of the large scale structures. At each stage, it has been connected to some observations.

The main paradigm lies on the dynamics of a universe that cools while expanding, the quantum fluctuations of a scalar field in the early universe, gravitational instability and then a progressive structuration that depends on the value of the cosmological parameters, the nature of the dark matter and the dissipative processes that affect baryons. Among the main successes of the model, we can recall: the origin of the variety of atoms (during BBN), the origin of matter (during reheating), the origin of the large scale structure (during inflation). This description is robust to our knowledge of the laws of physics since the energies involved since 1 s after the big-bang are smaller than 100 MeV, that is energies well-tested in the laboratory. It relies on nothing speculative and is an almost direct consequence of general relativity and known laboratory physics. We can state that the main uncertainty lies in the cosmological hypotheses. Only at earlier times are there doubts about the extrapolations used. This robustness is balanced by all the limitations we have discussed (one observable universe, observed from one point, the existence of horizons, the need for abduction, etc. ...)

The matter contained in the model includes photons (0.01%), baryons and electrons (5%), neutrinos (0.1-2%), non-baryonic dark matter (25%), a cosmological constant (or dark energy) (70%). This means that the minimal number of cosmological parameters boils down to 6: Ω_{baryon}, Ω_γ, Ω_ν, Ω_{cdm}, Ω_Λ (which reduces to 3 since Ω_γ, Ω_ν are determined and the sum of all Ωs is 1), the Hubble constant H_0, the amplitude A_S and spectral index n_S of the initial spectrum and the reionisation parameter τ_e. It can be trivially extended to include a spatial curvature (Ω_K), a dynamical dark energy component ($w, \ldots$) and extra-neutrinos (N_ν). These 6 parameters are today well measured and all the extra-parameters are well constrained.

The model nevertheless suffers from a series of problems and questions.

- *The lithium problem*, discussed in § 2.3.2.
- *The understanding of the form in which baryons are.* The baryonic density is well determined from CMB analysis but the baryon budget at low redshift is more difficult to establish: intergalactic gas (cold or hot), stars, planets...
- *The understanding of the end of the dark age*: formation of the first stars, origin of reionization, history of galaxies. Do we see all galaxies (in particular are there low brightness galaxies) and to which extent is the distribution of the galaxies a good tracer of the distribution of matter in the universe? What is the role of black holes in the formation of galaxies?...
- *Validity and precision of the perturbation theory.* See § 2.4.2.
- *The understanding of the nature and properties of dark matter.* This component of matter is not included in the model of particle physics as we know it but there exist many candidates in various of its extension (such as the lightest supersymmetric particle). Cosmology hence provides a strong indication for the need of an extension of the standard model of particle physics. It can be optimistically thought that the conjunction of astrophysical observations, direct searches and accelerator experiments shall characterize this particle within the coming ten years. An alternative to dark matter is the formulation of modified theories for gravity, among which MOND, but this route is not favored...
- *The cosmological constant problem or the understanding of the nature of dark energy.* This component is required to explain the late time acceleration of the universe. Dark energy confronts us with a compatibility problem since, in order to "save the phenomena" of the observations, we have to include new ingredients (constant, matter fields or interactions) beyond those of our established physical theories [11]. The important conclusion here is that we either have to accept a cosmological constant, as favored by all observations, or introduce new degrees of freedom in the model, either as geometric degrees of freedom of the cosmological solution, or as fundamental degrees of freedom (i.e., new physical fields), that can act as a new matter component or as a modification of general relativity if they are responsible of a long-range interaction due to their coupling with standard matter. See, e.g., Ref. [130] for the observational prospective on the understanding of dark energy.
- *The microphysics behind inflation.* For example (but far fom being limitative), To which extent can we reconstruct the potential of the inflaton from cosmological observations? How is the inflaton connected to the standard model of particle physics (can it be the Higgs [43], is single field inflation stable to the existence of other heavier fields [131])? Can one construct models without scalar field? How robust are the predictions and can one construct radically different models (as the now-ruled out opponent theory of topological defects)? Can one relate inflation to supersymmetry or string theory in a satisfying way? What happens to other fields? Should we be worried

by trans-Planckian modes? How robust is the hypothesis of a Bunch-Davies vacuum [101]? Can one detect features (non-Gaussianity, correlations) of the initial conditions that would reveal the physics at work? Can one constrain the number of e-folds? Can one detect the primordial gravity waves? What is the structure of spacetime in eternal inflation?...

- *Connecting inflation to the hot big-bang.* The physics of reheating needs to be better investigated, in particular to include the standard model of particle physics and the transition from the quantum perturbation to classical stochastic perturbations also needs to be better understood.

- *The backreaction question* discussed in § 2.6.

- *The effect of small scale structure* on the observation of the universe and to which extent it biases our observations; see § 2.5.4.

- *Emergence of complexity* The fact that we can understand the universe and its laws has also a strong implication on the structure of the physical theories. At each step in our construction of physical theories, we have been dealing with phenomena below a typical energy scale, for technological constraints, and it turned out (experimentally) that we have always been able to design a consistent theory valid in such a restricted regime. This is not expected in general and is deeply rooted in the mathematical structure of the theories that describe nature. We can call such a property a *scale decoupling principle* and it refers to the fact that there exist energy scales below which effective theories are sufficient to understand a set of physical phenomena that can be observed. *Effective theories* are the most fundamental concepts in the scientific approach to the understanding of nature and they always come with a domain of validity inside which they are efficient to describe all related phenomena. They are a successful explanation at a given level of complexity based on concepts of that particular levels. This implies that the structure of the theories are such that there is a kind of stability and independence of higher levels with respect to more fundamental ones. This lead to the idea of a hierarchy of physical theories of increasing complexity [132], which affects our understanding of causality to take into account contextuality, selection effects (in the Darwinian sense), emergence etc.

- *Fine tuning.* A related issues lies in the observational fact that we (as a complex biological system) are here observing the universe. It means that the fundamental laws of physics, which set the space of possibilities of higher complexity theories (such as chemistry, biology etc.) need to allow for the existence of complexity. The study of the variation of the fundamental constants have taught us that small variation in the fine structure constant or other parameters can forbid the emergence of such a complexity, one of the more critical parameter being the cosmological constant. How is our universe fine-tuned (and according to which measure)? is an essential question. It can be answered for some physical system, e.g., the production of carbon-12 in population III stars require a tuning at the level of 10^{-3} [133]. This line of arguments is an essential motivation to promote the idea of a multiverse. It still requires a better understanding.

- *Primordial singularity.* The cosmological model exhibits a spacelike singularity when extrapolated in the past. This big-bang is not a physical event but a limit in a given theoretical model. It is important to understand how this conclusion may be changed by quantum theories of gravity. It also appears as an explicit limit of our model that motivates many questions on the origin of the universe, that goes beyond the purpose of the model itself.
- *Links with quantum theories of gravity.* In the early universe quantum gravity may be important to understand the dynamics of the universe. A large activity focuses on the development of a realistic phenomenology of these theories, and in particular string theory.

Whatever the problem or the discrepancy in the model, one can always look for 3 kinds of solutions: astrophysical (i.e., an astrophysical effect that would affect and alter the observation), cosmological (i.e., the fact that the cosmological hypotheses are two strong and need to be relaxed, hence modifying our interpretation of the data) or physical (i.e., a need to modify the fundamental laws of physics).

An important step of the past decade has been the development of many tests of the hypothesis of the model, to be contrasted with the mainstream quest for a better precision of the measurement of the cosmological parameters. This includes tests of the field equations of general relativity on large scale [3, 11, 14, 15], tests of the equivalence principle by the constancy of fundamental constants [17, 18] (but also tests of some well-defined extensions of general relativity such as the scalar-tensor theories [67, 134]), tests of the Copernican principle (both homogeneity [21] and isotropy [101]), tests of the distance duality relation [82], and constraints on the topology of the universe [24, 25, 26]. No deviations from the standard model, despite the need of a dark sector, have been detected so far.

The field of cosmology can be seen as following two routes, mostly independent.

Observational cosmology has been booming, with the development of new techniques and of large surveys in all possible wavelengths. This has led to the idea of *precision cosmology*. One has to stress that measuring with higher accuracy does not mean that we get a better understanding of our model, and in particular one needs to be aware that the definition of the cosmological parameters depends on the cosmological model. One also needs to control the validity of the model and of its predictions with a similar accuracy. As discussed above, the precision of the observations may lead to question the validity of the use of a FL spacetime on all scales [42]. Again, this justify the need to test the hypothesis of the model. A trend has been to introduce arbitrary (or badly motivated) extensions of the model (and thus some extra-cosmological parameters) and to constrain them, backed up by the development of computational techniques. While it gives an idea of the validity of the minimal model, it does not teach much on physics. Note also that the interpretation of data also requires some deep investigation on the understanding of the non-linear regime, in particular by including all relativistic effects, and on the relation between perturbation theory and N-body simulations, which are essentially Newtonian. The precision on some parameters (such as the curvature parameter, the spectral index or its running) are indeed important while a high precision on the value of the cosmological constant does not improve our understanding of the physics behind.

Primordial cosmology tends to obtain a better description of the early phases of our the universe. It deals with many extension of the laws of nature as we know them. It is the playground of phenomenology for theories of quantum gravity and of many speculations, often out of reach of any observational testability, such as the multiverse. The important question is the "realism" of this phenomenology, which is often based on simplified version of the underlying theories. The guide here is mostly the search for a consistent picture and this is where one needs to be careful with the distinction between cosmology and Cosmology.

One century after the formulation of general relativity and ninety-eight years after the first relativistic cosmological model, cosmology has made significant progresses and managed to connect theory with a blooming observational activity. It offers new views of our universe.

Today, we can state that there is no need, from an observational and phenomenological point of view, to doubt general relativity (including a cosmological constant). Indeed the main problems here are theoretical (cosmological constant problem and the need for quantum gravity at high energy) but we may question that there exist some intermediate scale at which general relativity needs to be modified in a way that it would imprint phenomena in our observable universe. The trend to introduce often badly motivated extension of general relativity (which in particular do not solve the issue of the cosmological constant or of the quantum aspects) has mostly been induced by the need of a phenomenology for comparing to observations. This has indeed taught us the effects of many of these extensions, but they have to be considered only as toy models.

The next step in our understanding of gravity and of our universe lies probably in our ability to detect gravitational waves, which will open a new observational window on our universe, and inevitably lead to a new picture.

Acknowledgements

This work made in the ILP LABEX (under reference ANR-10-LABX-63) was supported by French state funds managed by the ANR within the Investissements d'Avenir programme under reference ANR-11-IDEX-0004-02. I thank George Ellis, Pierre Fleury and Cyril Pitrou for their comments.

References

[1] Ellis, G.F.R.: in *Handbook in Philosophy of Physics*. Ed. J. Butterfield and J. Earman (Elsevier, 2006), [arXiv:astro-ph/0602280].

[2] Uzan, J.-P.: in *Philosophy and Foundations of Physics*. (Cambridge University Press, 2016).

[3] Uzan, J.-P.: in *Dark Energy, observational and Theoretical approaches*, P. Ruiz-Lapuente Ed., (Cambridge Univ. Press, 2010), pp. 3-47, [arXiv:0912.5452].

[4] Ellis, G.F.R.: Q. Jl. astr. Soc. **16**, 245 (1975).

[5] Eisenstaedt, J.: in *Foundations of big-bang cosmology*, Coord. F.W. Meyerstein, (World Scientific, 1989).

[6] Ellis, G.F.R., Nel, S.D., Maartens, R., Stoeger, W.R., and Whitman, A.P.: Phys. Rep. **124**, (1985).

[7] Dunsby, P., et al.: JCAP **1006**, 017 (2010).

[8] Stephani, H., et al.: *Exact Solutions of Einstein's Field Equations* (Cambridge University Press, 2009); A. Krazinski, *Inhomogeneous Cosmological Models* (Cambridge University Press, 1997).

[9] Einstein, A.: Sitzungsber. K. Preuss. Akad. Wiss. 142 (1917).

[10] Duhem, P.: *Sozein ta phainomena: Essai sur la notion de théorie physique de Platon à Galilée* (Vrin, Paris, 1908); translated as *Sozein ta phainomena: an essay on the idea of physical theory from Plato ti Galileo*, (Univ. of Chicago Press).

[11] Uzan, J.-P.: Gen. Rel. Grav. **39**, 307 (2007).

[12] Will, C.M.: *Theory and experiment in gravitational physics*. (Cambridge University Press, 1981).

[13] Deruelle, N., and Uzan, J.-P.: *Théories de la relativité*. (Berlin, Paris, 2014).

[14] Uzan, J.-P.: Gen. Rel. Grav. **42**, 2219 (2010).

[15] Uzan, J.-P., and Bernardeau, F.: Phys. Rev. D **64**, 083004 (2001).

[16] Damour, T., and Nordtvedt, K.: Phys. Rev. Lett. **70**, 2217 (1993).

[17] Uzan, J.-P.: Rev. Mod. Phys. **75**, 403 (2003); Uzan, J.-P.: AIP Conf. Proc. **736**, 3 (2005); Uzan, J.-P.: Space Sci. Rev. **148**, 249 (2010).

[18] Uzan, J.-P.: Living Rev. Rel. **14**, 2 (2011).

[19] Peter, P., and Uzan, J.-P.: *Primordial Cosmology*. (Oxford Univ. Press, 2010).

[20] Ellis, G.F.R., and Stoeger, W.R.: Class. Quant. Grav. **4**, 1697 (1987).

[21] Uzan, J.-P., Clarkson, C., and Ellis, G.F.R.: Phys. Rev. Lett. **100**, 191303 (2008); Clarkson, C., Basset, B., and Hui-Ching Lu, T.: Phys. Rev. Lett. **101**, 011301 (2008); Caldwel, R.R., and Stebbins, A.: Phys. Rev. Lett. **100**, 191302 (2008); Zhang, P., and Stebbins, A.: Phys. Rev. Lett. **107**, 041301 (2011).

[22] Ellis, G.F.R., and Uzan, J.-P.: CRAS (2015), in press.

[23] Lachieze-Rey, M., and Luminet, J.-P.: Phys. Rept. **254**, 135 (1995); Uzan, J.-P., Lehoucq, R., and Luminet, J.-P.: Nucl. Phys. Proc. Suppl. **80** 0425 (2000).

[24] Weeks, J., Lehoucq, R., and Uzan, J.-P.: Class. Quant. Grav. **20**, 1529 (2003); Gausmann, E., et al.: Class. Quant. Grav. **18**, 5155 (2001); Uzan, J.-P., Lehoucq, R., and Luminet, J.-P.: Astron. Astrophys. **351**, 766 (1999); Lehoucq, R., Uzan,

J.-P., and Luminet, J.-P.: Astron. Astrophys. **363**, 1 (2000); Lehoucq, R., Luminet, J.-P., and Uzan, J.-P.: Astron. Astrophys. **344**, 735 (1999).

[25] Riazuelo, A., *et al.*: Phys. Rev. D **69** 103518 (2004); Riazuelo, A., *et al.*: Phys. Rev. D **69** 103514 (2004); Uzan, J.-P., *et al.*: Phys. Rev. D **69** 043003 (2004); Luminet, J.-P., *et al.*: Nature (London) **425** 593 (2003).

[26] Shapiro, J., et al.: Phys. Rev. D **75**, 084034 (2007); Fabre, O., Prunet, S., and Uzan, J.-P.: Phys. Rev. D **92**, 043003 (2015).

[27] Lemaître, G.: Ann. de la Soc. Sci. Bruxelles A **47**, 49 (1927). Published in translation in Month. Not. R. Astron. Soc. **91**, 483 (1931). Reprinted Gen. Rel. Grav. **45**, 1635 (2013); Lemaître, G.: Ann. Soc. Sci. Bruxelles A **53**, 51 (1933). Reprinted Gen. Rel. Grav. **29**, 641 (1997); Lemaître, G.: Proc. Natl. Acad. Sci. **20**, 1217 (1934).

[28] Friedmann, A.: Zeichrif. Phys. **10**, 377 (1922). Reprinted Gen. Rel. Grav. **31**, 1991 (1999); Friedmann, A.: Zeichrif. Phys. **21**, 326 (1924). Reprinted Gen. Rel. Grav. **3**, 2001 (1999).

[29] Hubble, E.: Proc. Natl. Acad. Sci. **15**, 168 (1929).

[30] Tolman, R.C.: *Relativity, Thermodynamics, Cosmology.* (Oxford University Press, Oxford 1934).

[31] Alpher, R.A., Bethe, H., and Gamow, G.: Phys. Rev. Lett. **73**, 803 (1948); Gamow, G.: Nature **162**, 680 (1948); Alpher, R.A., and Herman, R.: Nature **162**, 774 (1948); Gamow, G.: Phys. Rev. **74**, 505 (1948).

[32] Guth, A. H.: Phys. Rev. D **23**, 34756 (1981).

[33] Lifshitz, E. M.: J. Phys. (USSR) **10**, 116 (1946). Translated J. Phys. (1) **2**, 116 (1946).

[34] Harrison, E.R.: Rev. Mod. Phys. **39**, 862 (1967); Hawking, S.W.: Astrophys. J. **145**, 544 (1966); Weinberg, S.W.: *Gravitation and Cosmology: Principles and Applications of the General Theory of Relativity.* (Addison Wesley & son, 1972).

[35] Bardeen, J.: Phys. Rev. D **22**, 1882 (1980).

[36] Stewart, J.: Class. Quant. Grav. **7**, 1169 (1990).

[37] Mukhanov, V., and Chibisov, G.: JETP Lett. **33**, 532, (1981).

[38] Sachs, R.K., and Wolfe, A.M.: Astrophys. J. **147**, 73 (1967); Peebles, P.J.E., and Yu, J.T.: Astrophys. J. **162**, 815 (1970); Sunyaev, R.A., and Zel'dovich, Ya. B.: Astrophys. Space Science **7**, 3 (1970).

[39] Ostriker, J.P., and Steinhardt, P.J.: Nature **377**, 600 (1995).

[40] Perlmutter, S., *et al.*: Astrophys. J. **517**, 565 (1999); Riess, A.G., *et al.*: Astron. J. **116**, 1009, (1998).

[41] Fleury, P., Dupuy, H., and Uzan, J.-P.: Phys. Rev. D **87**, 123526 (2013).

[42] Fleury, P., Dupuy, H., and Uzan, J.-P.: Phys. Rev. Lett. **111**, 091302 (2013).

[43] Ellis, G.F.R., and Uzan, J.-P.: Astron. and Geophys. **55**, 1.19 (2014).

[44] Hawking, S.W., and Ellis, G.F.R.: *The Large Scale Structure of spacetime.* (Cambridge University Press, Cambridge, 1973).

[45] Betoule, M., et al.: Astron. Astrophys. **568**, A22 (2014).

[46] Bianchi, L.: Soc. Ital. Sci. Mem. Mat. **11**, 267 (1898); Ellis, G.F.R., and MacCallum, M.A.H.: Comm. Math. Phys. **12**, 108 (1969).

[47] Lemaître, G.: Ann. Soc. Sci. Bruxelles **A53** (1933); Tolman, R.C.: Proc. Nat. Acad. Sci. USA, **20**, 169 (1934); Bondi, H.: Mon. Not. R. Astr. Soc. **107**, 410 (1947).

[48] Lemaître, G.: Publications du Laboratoire d'Astronomie et de Géodesie de l'Université de Louvain **8**, 101–120 (1931).

[49] Bernstein, J.: *Kinetic theory in the expanding universe.* (Cambridge University Press, 1988); Stewart, J.: Lect. Notes in Phys. **10** (Springer, Berlin, 1971).

[50] Peebles, P.J.E.: Astrophys. J. **146**, 542 (1966).

[51] Bernstein, J., Brown, L.S., and Feinberg, G.: Rev. Mod. Phys. **61**, 25 (1989).

[52] Coc, A., Goriely, S., Xu, Y., Saimpert, M., and Vangioni, E.: Astrophys. J. **744**, 158 (2012).

[53] Coc, A., Uzan, J.-P., and Vangioni, E.: [arXiv:1307.6955].

[54] Coc, A., Uzan, J.-P., and Vangioni, E.: JCAP **1410**, 050 (2014).

[55] Pettini, M., and Cooke, M.: Month. Not. Astron. Soc. **425**, 2477 (2012).

[56] Cooke, R., Pettini, M., Jorgenson, R.A., Murphy, M.T., and Steidel, C.C.: Astrophys. J. **781**, 31 (2014).

[57] Aver, E., Olive, K.A., and Skillman, E.D.: JCAP **04**, 04 (2012); Aver, E., Olive, K.A., and Skillman, E.D.: JCAP **11**, 17 (2013).

[58] Rood, T. Bania R., and Balser, D.: Nature **415**, 54 (2002).

[59] Vangioni-Flam, E., *et al.*: Astrophys. J. **585**, 611 (2003).

[60] Spite, F., and Spite, M.: Astron. Astrophys. **115**, 357 (1982).

[61] Iocco, F., Bonifacio, P., and Vangioni, E.: Proceedings of the workshop *Lithium in the cosmos.* Mem. S. A. It. Suppl., 3 **22** (2012).

[62] Coc, A. , Pospelov, M., Uzan, J.-P., and Vangioni, E.: Phys. Rev. D **90**, 085018 (2014); Coc, B., Uzan, J.-P., and Vangioni, E.: Phys. Rev. D **87**, 123530 (2013).

[63] Sbordone, L., et al.: Astron. Astrophys. **522**, 26 (2010).

[64] Charbonnel, C., et al.: IAU Symposium **268**, *Light Elements in the Universe.* (Cambridge University Press, 2010).

[65] Frebel, A., and Norris, J.E.: *Published in 'Planets, stars and stellar systems',* *Springer.* Edts T. Oswalt & G. Gilmore **vol. 5** (2013) p. 55.

[66] Coc, A., et al.: Phys. Rev. D **76**, 023511 (2007); Coc, A., et al.: Phys. Rev. D **86**, 043529 (2012).

[67] Damour, T., and Pichon, B.: Phys. Rev. D **59**, 123502 (1999); Coc, A., et al.: Phys. Rev. D **73**, 083525 (2006); Coc, A., et al.: Phys.Rev. D **79**, 103512 (2009).

[68] Mather, J.C.: Astrophys. J. **512**, 511 (1999).

[69] Srianand, R., Petitjean, P., and Ledoux, C.: Nature **408**, 931 (2000).

[70] Tegmark, M., et al.: Phys. Rev. D **74**, 123507 (2006).

[71] Delubac, T., et al.: Astron. Astrophys. **574**, A59 (2015).

[72] Planck Collaboration, Adam, R., et al.: [arXiv:1502.0158].

[73] Clarkson, C., *et al.*: Month. Not. R. Astron. Soc. **426**, 1121 (2012).

[74] Baumann, D., Nicolis, A., Senatore, L., and Zaldarriaga, M.: [arXiv:1004.2488].

[75] Zwicky, F.: Helv. Phys. Acta **6**, 110 (1933).

[76] Babcock, H.W.: Lick Obs. Bull. **19**, 41 (1939).

[77] Ostriker, J.P., Peebles, P.J.E., and Yahil, A.: Astrophys. J. Lett. **193**, L1 (1974).

[78] Einsasto, J., Saar, E., Kaasik, A., and Chernin, A.D.: Nature **252**, 111 (1974).

[79] Ostriker, J.P., and Peebles, P.J.E.: Astrophys. J. **186**, 467 (1973).

[80] White, S.D.M., and Rees, M.J.: Month. Not. R. Astron. Soc. **183**, 341 (1978).

[81] Press, W.H., and Schechter, P.: Astrophys. J. **187**, 425 (1974).

[82] Uzan, J.-P., Aghanim, N., and Mellier, Y.: Phys. Rev. D **70**, 083533 (2004); Ellis, G., Poltis, R., Uzan, J.-P., and Weltman, A.: Phys. Rev. D **87**, 06520 (2013).

[83] Deffayet, C., Harari, D., Uzan, J.-P., and Zaldarriaga, M.: Phys. Rev. D **66**, 043517 (2002).

[84] Bartelmann, M., and Schneider, P.: Phys. Rep. **340**, 291 (2001).

[85] Mellier, Y.: Annu. Rev. Astron. Astrophys. **37**, 127 (1999).

[86] Wittman, D.: Lec. Notes Phys. **608**, 55 (2002).

[87] Sachs, R.K., and Wolfe, A.M.: Astrophys. J. **147**, 73 (1967).

[88] Panek, M.: Phys. Rev. D **34**, 416 (1986).

[89] Hu, W., and Sugyiama, N.: Phys. Rev. D **51**, 2599 (1995); Hu, W., and Sugyiama, N.: Astrophys. J. **444**, 489 (1995); Sakharov, A.: Zh. Eksp. Teor. Fiz. **49**, 345 (1965), [*Sov. Phys. JETP* **22**, 241 (1966)]; Silk, J.: Astrophys. J. **151**, 459 (1968).

[90] Bond, J.R., and Efstathiou, G.: Astrophys. J. Lett. **285**, L45 (1984); Ma, C.P., and Bertschinger, E.: Astrophys. J. **455**, 7 (1995); Hu, W., and White, M.: Phys. Rev. D **56**, 596 (1997).

[91] Pitrou, C.: Class. Quant. Grav. **24**, 6127 (2007); Pitrou, C., Uzan, J.-P., and Bernardeau, F.: Phys. Rev. D **78**, 063526 (2008); Pitrou, C.: Class. Quant. Grav. **26**, 065006 (2009); Pitrou, C.: Gen. Rel. Grav. **41**, 2587 (2009).

[92] Planck Collaboration: Ade, P.A.R., et al.: [arXiv:1303.5077].

[93] Futamase, T., and Sasaki, M.: Phys. Rev. D **40**, 2502 (1989); Cooray, A., Holtz, D., and Huterer, D.: Astrophys. J. **637**, L77 (2006); Dodelson, S., and Vallinotto, A.: Phys. Rev. D **74**, 063515 (2006).

[94] Valageas, P.: Astron. Astrophys. **354**, 767 (2000); Bonvin, C., Durrer, R., and Gasparini, M.: Phys. Rev. D **73**, 023523 (2006); Meures, N., and Bruni, M.: Mon. Not. Roy. Astron. Soc. **1937** (2012).

[95] Einstein, A., and Straus, E.G.: Rev. Mod. Phys. **17**, 120 (1945); Einstein, A., and Straus, E.G.: Rev. Mod. Phys. **18**, 148 (1945).

[96] Zel'dovich, Y.B.: Sov. Astron. Lett. **8**, 13 (1964).

[97] Feynman, R.P.: Unpublished colloquium given at the California Institute of Technology (1964).

[98] Dashevskii, V.M., and Zel'dovich, Y.B.: Sov. Astronom. **8**, 854 (1965); Bertotti, B.: Royal Society of London Proceedings Series A **294**, 195 (1966); Gunn, J.E.: Astrophys. J. **150**, 737 (1967); Gunn, J.E.: Astrophys. J. **147**, 61 (1967); Weinberg, S.: Astrophys. J. **208**, L1 (1973).

[99] Fleury, P., Larena, J., and Uzan, J.-P.: JCAP (in press), [arXiv:1508.07903].

[100] Pereira, T., Pitrou, C., and Uzan, J.-P.: JCAP **09**, 006 (2007); Pitrou, C., Pereira, T., and Uzan, J.-P.: JCAP **04**, 004 (2008).

[101] Pitrou, C., Uzan, J.-P., and Pereira, T.S.: Phys. Rev. D **87**, 043003 (2013); Pitrou, C., Pereira, T.S., and Uzan, J.-P.: Phys. Rev. D **92**, 023501 (2015); Pereira, T.S., Pitrou, C., and Uzan, J.-P.: [arXiv:1503.01127]; Fleury, P., Pitrou, C., and Uzan, J.-P.: Phys. Rev. D **91**, 043511 (2015).

[102] Buchert, T.: Gen. Rel. Grav. **32** (2000) 105; Buchert, T.: Gen. Rel. Grav. **33** (2001) 1381; Buchert, T., et al.: [arXiv:1505.0780]; Green, S. R., and Wald, R. M.: Class. Quant. Gravi. **31**, 234003 (2014).

[103] Carlson, J., White, M., and Padmanabhan, N.: Phys. Rev. D **80**, 043531 (2009).

[104] Adamek, J., Daverio, D., Durrer, R., and Kunz, M.: [arXiv:1509.01699].

[105] Gliner, E.: Sov. Phys. JETP **22**, 378 (1966).

[106] Brout, R., Englert, F., and Gunzig, E.: Ann. Phys. **115**, 78 (1978).

[107] Starobinsky, A.: JETP Lett. **30**, 682 (1979).

[108] Linde, A. D.: Phys. Lett. B **108**, 389 (1982).

[109] Albrecht, A., and Steinhardt, P.J.: Phys. Rev. Lett. **48**, 1220 (1982).

[110] Linde, A.: *Particle physics and inflationary cosmology.* (Harwood Academic Publishers, 1990).

[111] Linde, A.: Phys. Lett. B **129**, 177 (1983).

[112] Kofman, L.: in *Particle Physics and Cosmology: the Fabric of Spacetime.* Les Houches 2006, PP. 195, F. Bernardeau et al. Eds. (Elsevier, 2007).

[113] Gibbons, G., and Hawking, S.: Phys. Rev. D **15**, 2738 (1977); Bunch, T., and Davies, P.: Proc. Roy. Soc. A **360**, 117 (1978).

[114] Susskind, L.: [arXiv:hep-th/0302219].

[115] Mukhanov, V., Feldman, H., and Brandenberger, R.: Phys. Rept. **215**, 203 (1992).

[116] Vilenkin, A., and Ford, L.: Phys. Rev. D **26**, 1231 (1982); Linde, A.: Phys. Lett. B **116**, 335 (1982).

[117] Hawking, S.: Phys. Lett. B **115**, 295 (1982); Starobinsky, A.: Phys. Lett. B **127**, 175 (1982); Guth, A., and Pi, S.-Y.: Phys. Rev. Lett. **49**, 1110 (1982); Bardeen, J., Steinhardt, P., and Turner, M.: Phys. Rev. D **28**, 679 (1983).

[118] Shtanov, Y., Traschen, J., and Brandenberger, R.: Phys. Rev. D **51**, 5438 (1995); Kofman, L., Linde, A., and Starobinsky, A.: Phys. Rev. D **56**, 3258 (1997); Greene, P., et al.: Phys. Rev. D **56**, 6175 (1997).

[119] Lidsey, J., et al.: Rev. Mod. Phys. **69**, 373 (1997).

[120] Uzan, J.-P., Kirchner, U., and Ellis, G.F.R.: Month. Not. R. Astron. Soc. **344**, L65 (2003).

[121] Pereira, T., Pitrou, C., and Uzan, J.-P.: JCAP **09**, 006 (2007); Pitrou, C., Pereira, T., and Uzan, J.-P.: JCAP **04**, 004 (2008).

[122] Grib, A., Mamaev, S., and Mostepanenko, V.: *Quantum effects in strong external fields.* (Atomizdat, Moscou, 1980).

[123] Mukhanov, V., and Winitzki, S.: *Introduction to Quantum Effects in Gravity.* (Cambridge University Press, 2007).

[124] Birell, N., and Davies, P.: *Quantum fields in curved space.* (Cambridge University Press, 1982).

[125] Planck Collaboration: Ade, P. A. R., et al.: [arXiv:1502.02114].

[126] Bernardeau, F., Kofman, L., and Uzan, J.-P.: Phys. Rev. D **70**, 083004 (2004).

[127] Kleban, M.: Class. Quant. Grav. **28**, 204008 (2011); Chang, S., Kleban, M., and Levi, T. S.: JCAP **0904**, 025 (2009).

[128] Vilenkin, A.: Phys. Rev. D **27**, 2848 (1983); Linde, A.: Phys. Lett. B **175**, 395 (1986); Goncharov, A.S., Linde, A.D., and Mukhanov, V.F.: Int. J. Mod. Phys. A **2**, 561 (1987).

[129] Guth, A.: Phys. Rept. **333**, 555 (2000).

[130] Jain, B., et al.: [arXiv:1309.5389].

[131] Renaux-Petel, S., and Turzynski, K.: [arXiv:1510.01281].

[132] Ellis, G.F.R.: Nature **435**, 743 (2005); Ellis, G.F.R.: "On the nature of causation in complex systems"; Ellis, G.F.R.: "Top-down causation and emergence: some comments on mechanisms"; Ellis, G.F.R.: "Physics and the real world"; Uzan, J.-P.: in *The Causal universe*. (Copernicus center press, 2013), p. 93.

[133] Ekström, S., et al.: Astron. Astrophys. **514**, A62 (2010).

[134] Martin, J., Schimd, C., and Uzan, J.-P.: Phys. Rev. Lett. **96** 061303 (2006).

Jean-Philippe Uzan
Institut d'Astrophysique de Paris
CNRS/UMR 7095
Université Pierre-et-Marie Curie Paris VI
98 bis, bd Arago
75014 Paris
France
e-mail: uzan@iap.fr

The Universe, 73–92

The Planck Mission and the Cosmic Microwave Background

Jean-Loup Puget

Abstract. Relevant observations to test cosmological models have increased dramatically in the last fifteen years. The *Planck* space mission for the observation of Cosmic Microwave Background has brought a large increase in the accuracy of the maps of its temperature and polarization. These results have confirmed in a spectacular way the ΛCDM model and determined the parameters attached to the present content and dynamics of our universe with sub percent accuracy. The combination with observations of the baryon acoustic oscillations measurements made this set of parameters even tighter. The predictions from the inflation paradigm of the early universe physics have been confirmed. The tensor-to-scalar modes ratio emerging from the early universe is only bounded by an upper limit. We show the robustness of the model by checking its assumptions (flat space geometry, gaussianity and adiabaticity of fluctuations). We also add to the ΛCDM six parameters key physical constants to the cosmological model that we constrain from the CMB anisotropies. This demonstrates the complementarity of the standard model of particle physics and what appears now to be the standard model of cosmology which together account for a very large body of high accuracy observations but need the introduction of new physics.

Keywords. Planck space mission, cosmic microwave background, lambda cold dark matter model, baryon acoustic oscillations, inflation, standard model of particle physics.

1. Introduction: historical perspective on cosmological theories and observations

The big bang theory described by J.-P. Uzan in this seminar has emerged in the first half of the twentieth century. It was made possible by general relativity as the

---

This paper presents the content of the seminar given on the 21st of November 2015 and the references to the *Planck* collaboration papers from which the results are taken.

theory of gravity and based on general principles trying to explore the simplest possible models and introduce more complicated models only when observations would impose it. The initial ones were that the universe is, on very large scales, homogeneous, isotropic and stationary.

The identification of galaxies similar to the Milky way and measurements of their distance and relative velocity allowed Hubble to show that the universe was expanding and to compute the first order of magnitude of the average density of matter in form of stars in the local universe. This forced the theoreticians to remove the stationarity hypothesis. The other observational parameter which was key to reinforce the theory of the so-called "big-bang" was the universal abundance of helium with respect to hydrogen in stars which led to the hypothesis of primordial nucleosynthesis of helium and the prediction of the Cosmic Microwave Background (CMB) by Alpher, Bethe, and Gamow [1] and subsequent work mostly done by Alpher. The discovery of an isotropic microwave background in 1965 by Penzias and Wilson [2] interpreted by Dicke et al. [3] to be the predicted CMB was a spectacular confirmation of a very daring theoretical prediction considering the state of observations when it was made. Such a pattern happened a number of times in cosmology: theoretical predictions are made, often on the basis of very general considerations of theoretical physics, which are tested and verified much later. We will show that this is still the case today.

The early observations of Hubble were on the content in baryonic matter and its dynamical behavior (the Hubble expansion) in the nearby universe. In the late 1960s, the big bang theory was nowhere near what you would expect from a standard model of cosmology. In the nineteen thirties, Zwicky had shown very early that galaxies where much more massive than the mass of stars seen in them [4, 5]. The "missing mass", which is now referred to as "dark matter", was a direct consequence of observations as it relied on well measured rotation velocities of stars and later of interstellar hydrogen in galaxies. It was also seen later in clusters of galaxies. Nevertheless the nature of the missing mass was not known and an origin in modifications of the law of gravity on large scale could not be excluded.

The prediction of the CMB rests on physics taking place in conditions extrapolated backwards in the history of expansion by a fantastic factor of a billion in the length scale and temperature of the universe from those observed by Hubble in the local universe. This nucleosynthesis had to happen only a few hundred seconds after the big bang. After the discovery of the CMB, it took 25 years before the technology for a cryogenically cooled instrument in space (COBE-FIRAS), permitted to Mather et al. [6] to check that the CMB spectral energy distribution had a very nearly Planckian spectrum with deviations from it less than 10^{-4}. This was the second spectacular verification of a cosmological model prediction done many years earlier.

It was also predicted very soon after the discovery of the CMB that adiabatic inhomogeneities in the energy density distribution could lead to the formation of all observed structures and would lead to acoustic oscillations during the radiation dominated phase of the expansion, and the harmonic modes of these at the time of recombination would leave acoustic peaks in the spatial distribution power spectrum observable on the CMB. It was shown by Zeldovich, Kurt, and Sunyaev [7] and by Peebles and Yu [8] that, in that case, these acoustic peaks would enable to measure the content of the universe with high accuracy.

Independantly, Harrison [9] and Zeldovich [10] separately remarked that the initial spectrum should be non divergent on either very large or very small scales and that the power law index of inhomogeneities should be close to a scale invariant one with the n_s parameter in models of the universe very close to one.

In the early seventies, the origin of the large scale structures was not understood as there was little evidence about an evolution of galaxies. The large scale geometry of the universe (Euclidean, closed or open) was also unknown, and the Hubble constant was known only within a factor of two uncertainty. For thirty years, there was no model which could account for the few basic cosmological observations coherently.

The development of the observed Large Scale Structures, especially clusters of galaxies and their number density, was requiring more mass than the baryonic mass observed in form of stars leading to the alternative of dark matter or to assume that most of the baryons in the universe was in diffuse gas. This was shown during this period to be the case in massive galaxy clusters: a hot gas radiating in X-rays dominates the baryon mass, but not enough to explain the dynamics of galaxies in clusters. Furthermore, there was no observational evidence for such a large amount of hot gas associated with field galaxies. The possible presence of a dark matter component was one way to solve this problem.

Another instrument on the COBE satellite (DMR) discovered the anisotropies in the CMB which were needed to explain the creation of structures in the universe from a nearly uniform distribution with very small primordial inhomogeneities growing under gravity [11]. The anisotropies detected on very large scale by COBE-DMR were too small if the baryonic matter was the only driver of the structure formation. Non-adiabatic inhomogeneities were considered, for example isocurvature [12]. This could be resolved by the introduction of a dark matter with very small coupling to the radiation and baryonic matter but which dominates the matter content.

2. Towards a standard model in cosmology

Around the year 2000, the picture changed dramatically with the first evidence from a specific type of supernovae (SNIa) that the universe expansion velocity was increasing at low redshifts instead of decreasing under the influence of the gravity of matter [13, 14]. Such observations are difficult, small statistics and they depend on understanding the possible SNIa luminosity evolution associated for example with the chemical evolution. Soon after, lensing observations led to the same conclusion. In cosmological models based on general relativity, this implied that at low redshift a cosmological constant term similar to the one introduced by Einstein to allow a stationary universe is present.

This could equivalently be that a large fraction of the stress-energy tensor is made of a dynamical component which has a negative pressure. If such a Λ term were introduced in the Einstein equation driving the cosmological expansion, the contradictions between the basic observations disappear and this was referred to as the "concordance model". This was still far from a standard model of cosmology analogous to the standard model of particle physics: the number of observations this model was explaining was small and the accuracy was low.

The ΛCDM introduced after 2000 is defined as a spatially flat geometry universe in which the energy density is dominated by a dark energy component (the Λ term: cosmological constant or a dynamical field with an equation of state $P = -E$) and the matter is dominated by a dark matter component (CDM part). The 2.725 K black body radiation completes the description of the present energy content of the universe. Furthermore, initial adiabatic density fluctuations with a power spectrum of spectral index (the Harrison-Zeldovich one) $n_s = 1$ is assumed to give rise to all structures observed now. The Thompson optical depth of free electrons from the present time to the time when the first sources reionized matter complete the ΛCDM parameter set. The physics is described by the standard model of particle physics and by general relativity for gravity. The cosmological model has only six additional free parameters to be deduced from observations. Combined with standard physics, the dynamics of the universe, from equations derived from general relativity, is set. This is due to the fact that the classical behavior in the expansion of these components is known and depends only of the equation of state (pressure vs energy density) for the three component of the energy density: negligible pressure, relativistic component $(P = 1/3E)$, dark energy $(P = -E)$. Dark matter and dark energy are two components not identified yet and beyond the standard model.

This present content of the universe is given by astrophysical observations but with a low precision: relativistic energy density (electromagnetic radiation and neutrinos), baryonic matter, dark matter and dark energy. Their associated parameters and the present expansion rate are needed for a full description of the cosmological evolution as long as the standard model of particles is still valid and that the additional hypothesis of the ΛCDM are understood and fulfilled. The goal of observational cosmology has been set to check the robustness of the model and to measure as accurately as possible all these parameters.

The SNIa observations give an estimate of the cosmological constant term, estimates of the baryonic matter density from big bang nucleosynthesis and of dark matter from structure formation observations (cluster density and lensing observations) and measurement of the present expansion velocity (the Hubble constant) by observations of distant galaxies. Although this introduction of this model was a breakthrough in cosmology, the parameters were not precisely set and strong a priori hypotheses (flat geometry, adiabatic gaussian density fluctuations) needed to be tested. The flat space geometry was challenging because, in a "classical big bang", the curvature increases with time. The initial conditions should thus have been extremely close to Euclidean in the early universe. Furthermore, we also need to find a physical source for the initial acceleration of the expansion of the primordial universe and the source of the adiabatic energy density fluctuations. In this model, the adiabatic fluctuations are imprinted as the CMB temperature and polarization anisotropies at a redshift of 1100 when the hydrogen in the universe recombines and let the radiation comes to us freely. We thus have a tool to observe the state of the universe at this redshift (380 000 years after the big bang) to complement the more local astrophysical observations. These fluctuations behave as as sound oscillations at times when they are within the horizon and are amplified by gravity when they are at larger scales. It was predicted after the discovery of the CMB that the power spectrum of the CMB anisotropies reflecting these acoustic modes would be a key observational tool for cosmology if they were present in the microwave sky [15, 8].

3. Observations of the Cosmic Microwave Background anisotropies

On angular scales larger than one degree, these anisotropies reflect the primordial spectrum. At scales smaller than the acoustic horizon they show acoustic peaks which depend on the cosmological parameters.

The search for the first acoustic peak was a big topic for observational cosmology in the 1990s as it was understood that its position in angular scale would reveal the geometry of the universe. Although marginal detections were done, it finally converged with observations carried out with two balloon borne experiments Boomerang [17] and Maxima [18] which confirmed spectacularly the flat space geometry of the universe on large scales assumed in ΛCDM. This was followed by the first year WMAP results [19] which confirmed the power of CMB anisotropies observations for high precision cosmology. WMAP was a passively cooled experiment at frequencies lower than the peak frequency of the CMB radiation. This experiment used radio type detection with newly developed high frequency HEMT amplifiers up to 94 GHz. This space mission was built within the short development time scale required by the NASA Explorer program.

Planck was selected at the same time as WMAP and was combining two instruments: one, the Low Frequency Instrument (LFI), was also HEMT based at frequencies comparable to WMAP, thus using established technologies but cryogenically cooled to 20 K. The other, the High Frequency Instrument (HFI), was using cryogenically cooled bolometers at 100 mK and thus more ambitious with better sensitivity and angular resolution but also more risky. The project suffered delays due to the Ariane 5 launcher first flight failure and was launched in 2009. It became the third generation space mission after COBE and WMAP with a stronger emphasis on the polarization of the CMB which is a much weaker signal. *Planck* had a number of new technologies: the space qualified dilution cooler was the main one. It was combined with a performant passive cooling and two other active coolers at 4 K and 20 K. This latter was also cooling the LFI HEMT amplifiers which required a high heat lift. This cooler was based on hydrogen sorption with no moving parts, leaving only two compressors for the 4K cooler which used a sophisticated vibration damping control to avoid inducing extra noise on the bolometers. Fig. 1 shows the HFI focal plane unit with its different cryogenic stages.

The HFI detector chains for the CMB frequencies (100 and 143 GHz) were mostly limited by the fundamental limit set by the photon noise of the cosmic background itself. The LFI and HFI combination was covering a broad range of frequencies (30 GHz to 1 THz) and thus controlling well the galactic and extragalactic foregrounds due to synchrotron emission on the low frequency side and dust emission on the high frequency side. The main CMB channels are concentrated near the minimum of the foregrounds at 70, 100, and 143 GHz.

The *Planck* satellite was orbiting the L2 Lagrange point of the sun-earth system which allows a very good rejection of the thermal radiation from the sun, the earth and the moon. It worked extremely well and provided results with the predicted goal sensitivity for 2.5 years for HFI (limited by the amount of ^{3}He carried for the dilution cooler) and 4 years for the LFI.

4. Determination of the ΛCDM cosmological model parameters

The CMB all sky map (Fig. 2) is extracted from the frequency maps using all the frequencies to remove the galactic and extragalactic foregrounds.

In the ΛCDM model, scalar primordial fluctuations are assumed Gaussian and thus all the information is contained in the power spectrum. The *Planck* measurement of these power spectra in temperature and E-modes polarization can be seen in Fig. 3. The parity invariant part of the polarization map is refereed as the E-modes and the part changing sign under parity the B-modes part by analogy with the electric and magnetic fields.

These power spectra are plotted as a function of the multipoles of the spherical harmonics decomposition of the maps after summing over the m parameter, and the plotted quantity is $D_\ell = C_\ell \times \ell \times (\ell+1)/2\pi$ for TT and TE. For EE, the C_ℓ is shown. In each figures, the bottom part shows the residuals between the data and the ΛCDM model with the *Planck* parameters. In temperature, the uncertainties are dominated by the cosmic variance of the signal measured, for multipoles less than about 1500. Above that value they are dominated by the noise. In polarization, the noise dominates for $\ell > 500$. Note that the units are μK^2 for spectra involving temperature but $10^{-5}\mu K^2$ for the E-modes power spectrum.

The accuracy of the determination of all the ΛCDM parameters is limited when using only the CMB power spectra because of the degeneracies between parameters, the main one being the geometric degeneracy. Figure 4 illustrates that point. The points represent the prediction of the model in the plane $\Omega_\Lambda - \Omega_m$, and are color coded as function of the Hubble parameter. The error contours given by

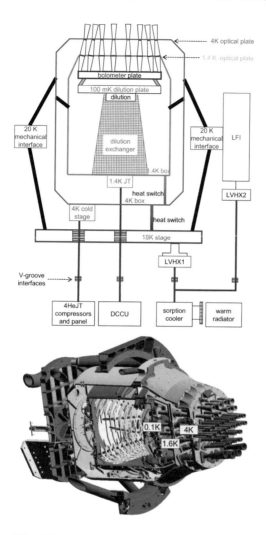

FIGURE 1. The *Planck* HFI instrument. Top : the thermal design schematics. Bottom: a view showing the 4K box holding the feed horns collecting the radiation from the telescope, the 1.6 K box holding the filters and the 100 mK plate holding the bolometers [16].

the black contours are from *Planck* temperature and polarization power spectra only. It can be seen that the CMB models points lie on a degeneracy zone close to the dotted line representing $\Omega_{tot} = \Omega_\Lambda + \Omega_m = 1$ which is the flat space geometry condition of ΛCDM. If this assumption was not made, low values of the Hubble constant are allowed outside the $\Omega_{tot} = 1$.

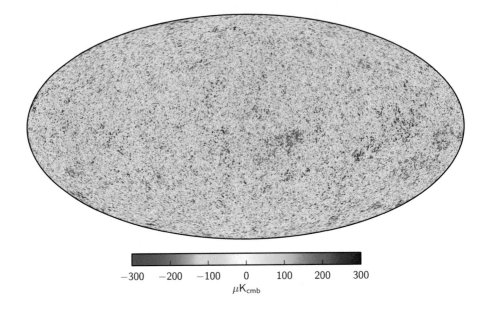

FIGURE 2. The *Planck* CMB temperature all sky map in galactic coordinates [20].

In the ΛCDM model the assumption on the geometry removes this degeneracy. This degeneracy prevents also CMB only data to measure the Hubble constant accurately. The sensitivity of *Planck* allows to measure the lensing of the CMB itself with enough accuracy (see Fig. 5) to remove mostly the geometric degeneracy when used in combination with the temperature and polarization power spectra. The blue patch, sitting nearly on the dotted line, shows the reduction of the errors brought by using jointly the CMB power spectra and lensing signals. Finally, the even smaller red patch shows the additional reduction of the uncertainties which is obtained by adding another geometric measurement on galaxies showing the Baryonic Acoustic Oscillations (BAO) reflecting at low redshift the acoustic peak seen at red shift of 1100 in the CMB improves the reduction of this degeneracy.

Using the flat space assumption was the condition for WMAP to get all the cosmological parameters. Using the value of the Hubble constant measured from measurements of radial velocity and distance of galaxies is another way. The measurement of the Hubble constant is a notoriously difficult one as it requires a chain of astronomy distance indicators going from the geometrical parallax of stars to the distance of galaxies which are more and more difficult when the galaxies are more distant. *Planck* is able to measure independently the Hubble constant at a value lower than the one measured on extragalactic sources although not incompatible with their uncertainties.

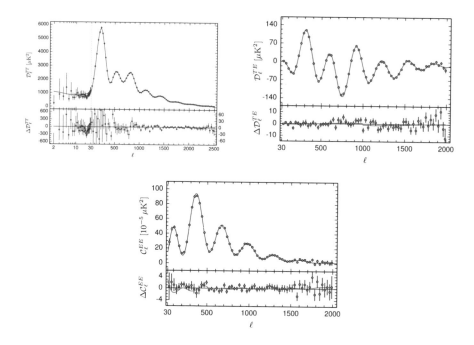

FIGURE 3. *Planck* CMB power spectra. The power spectra for temperature (left panel), temperature-E modes polarization cross spectrum (center panel), E-modes spectra (right panel). The residuals between observations and the *Planck* best model are shown in the bottom panels. (Ref. [21].)

This is a spectacular illustration of a trend shown in the last release of the *Planck* data. The addition of the polarization, the lensing and of other geometric measurements independent of astrophysical uncertainties comfort the cosmological model and reduces the uncertainties of the parameters within the ones obtained by the CMB temperature only data.

Table 1 illustrates the fact that the independent CMB polarization observations leads to the same cosmological parameter set and the improves the accuracy.

Figure 6 shows the excellent agreement between the prediction for the acoustic scale distance ratio from the *Planck* base ΛCDM model as a function of redshift and the BAO measurements. especially the best and most recent one (BOSS CMASS).

One parameter of the ΛCDM model was not measured accurately in WMAP nor on the first *Planck* release. This is the reionization parameter which measures the Thompson optical depth of the free electrons column density from the present

J.-L. Puget

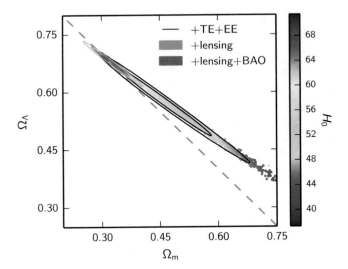

FIGURE 4. Constraints in the $\Omega_\Lambda - \Omega_m$ plane from the *Planck* 2015 release (black contours). The blue contours show the breaking of the geometrical degeneracy from the introduction of the lensing data and the blue contours when BAO measurements are introduced. (Ref. [21].)

Parameter	Planck TT+lowP	Planck TT,TE, EE+lowP
$\Omega_b h^2$	0.02222 ± 0.00023	0.02225 ± 0.00016
$\Omega_c h^2$	0.1197 ± 0.0022	0.1198 ± 0.0015
$100\theta_{MC}$	1.04085 ± 0.00047	1.04077 ± 0.00032
τ	0.078 ± 0.019	0.079 ± 0.017
$\ln(10^{10} A_s)$	3.089 ± 0.036	3.094 ± 0.034
n_s	0.9655 ± 0.0062	0.9645 ± 0.0049
H_0	67.31 ± 0.96	67.27 ± 0.66
Ω_m	0.315 ± 0.013	0.3156 ± 0.0091
σ_8	0.829 ± 0.014	0.831 ± 0.013
$10^9 A_s e^{-2\tau}$	1.880 ± 0.014	1.882 ± 0.012

TABLE 1. Parameters of the base ΛCDM model from the temperature only *Planck* data (left column) and combined with the polarization data. (Ref. [21].)

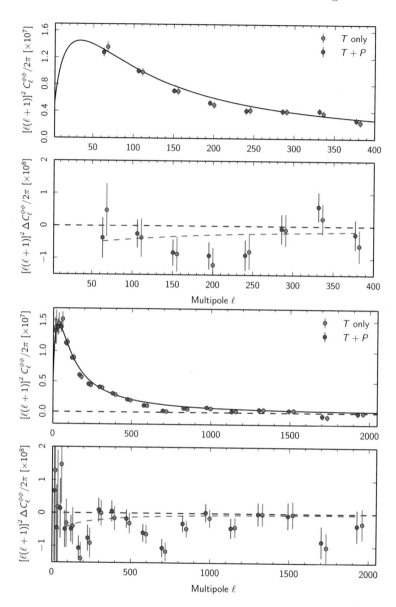

FIGURE 5. *Planck* measurements of the lensing power spectrum compared to predictions of the best fitting base ΛCDM model. The lower panels show the difference with the model. (Ref. [21].)

J.-L. Puget

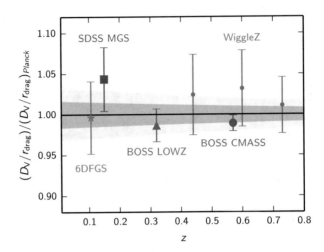

FIGURE 6. Acoustic-scale distance ratio in the base ΛCDM model
divided by the mean distance ratio from *Planck*. (Ref. [21].)

time to the time at which the first sources reionized the universe. The best measurement of this parameter is expected to be from the EE modes of the CMB
polarization data at very low multipoles ($\ell = 3 - 10$) which gives a measurement
almost independantly of the other parameters. Nevertheless, this measurement is
a very difficult one.

The τ value in Table 2 using the *Planck* data and the BAO measurements is
$\tau = 0.0066 \pm 0.012$. Using the *Planck EE* spectra at very low multipoles is now
possible, thanks to the understanding and removal of one non negligible systematic
effect. [23] finds

$$\tau = 0.055 \pm 0.009$$

which confirms and extend the trend towards lower values of τ shown in Fig. 7 and
the much smaller uncertainties showing for the first time a clear detection with
more than $6\,\sigma$.

5. Extensions of the base ΛCDM model

The extensions of the ΛCDM model can be of different kinds. Some extensions aim
at checking one by one the ΛCDM assumptions. We can also confront consistency
of predictions of the cosmological model with astrophysical measurements.

Big bang nucleosynthesis which led to the prediction of the existence of the
CMB is an obvious assumption to test. Big bang nucleosynthesis predictions of
helium and deuterium abundances as a function of baryon density can be compared
with astrophysical measurements of deuterium abundance which leads to a baryon

Parameter	TT,TE,EE+lowP+lensing+ext 68% limits
$\Omega_b h^2$	0.02230 ± 0.00014
$\Omega_c h^2$	0.1188 ± 0.0010
$100\theta_{MC}$	1.04093 ± 0.00030
τ	0.066 ± 0.012
$\ln(10^{10} A_s)$	3.064 ± 0.023
n_s	0.9667 ± 0.0040
H_0	67.74 ± 0.46
Ω_Λ	0.6911 ± 0.0062
Ω_m	0.3089 ± 0.0062

TABLE 2. The table illustrates the improvement brought by the introduction of observations data from galaxy surveys (BAO, JLA, H_0). (Ref. [21].)

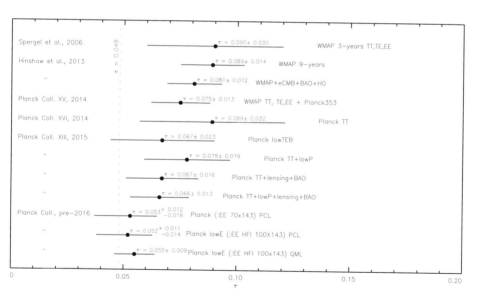

FIGURE 7. History of τ determination with WMAP and *Planck* (Ref. [23]).

density which is then compared to the *Planck* measurement. Taking the effective number of neutrino species from the standard model ($N_{eff} = 3.04$), there is full agreement with the best abundances astrophysical measurements. The accuracy of the *Planck* predictions is much better than the astrophysical measurements for helium.

The nucleosynthesis is affected by the number of species of neutrinos. The constraints on neutrino species from the *Planck* data is

$$N_{\text{eff}} = 3.04 \pm 0.18 \; (Planck \; \text{TT, TE, EE+low P+BAO}).$$

The CMB data also gives an upper limit on the sum of neutrino masses:

$$\sum m_{\nu} < 0.17 \; eV \, (Planck \; \text{TT, TE, EE+low P+BAO}).$$

The CMB temperature affects the recombination history and thus the acoustic peaks. One can ignore the COBE-FIRAS measurement of the CMB temperature (2.725 K) and adds to the parameters the temperature of present CMB as another parameter of the cosmological model to be fitted on the *Planck* anisotropy data. We find

$$T = 2.718 \pm 0.021 \; \text{K} \; (Planck \; \text{TT, TE, EE+low P+BAO}).$$

We can also assume that one important physical constant for cosmology is not known and to be extracted from the cosmological observations together with the cosmological parameters. The atomic physics rate of the forbidden transition 2s-1s of the hydrogen atom, which is important for recombination, can also be extracted from the CMB anisotropies *Planck* data as shown in Fig. 8 and is in agreement with the theoretical value.

A similar example is in the nuclear physics parameter critical for big bang nucleosynthesis. The nuclear rate $d(p, \gamma)^3He$ rate is controlling the deuterium abundance. We can extract a probability distribution for this rate compared to the rate used. We find 1.109 ± 0.058.

In conclusion, these examples show that we now have a standard cosmological model based on the CMB anisotropies measured with *Planck* combined with the BAO, JLA and H_0 observations which is encompassing not only a large number of cosmological observations but also is now constraining some of the underlying physics.

6. Beyond the Standard Model of particle physics

The dark energy equation of state in the ΛCDM model is the one given by a cosmological constant. We can open the parameters of the dark energy component w_0 and w_a. Figure 9 shows no evidence for deviation away from the cosmological constant values.

Some a-priori assumptions of the ΛCDM model are not driven by known physics and require physics beyond the standard model of particle physics: neither the flatness of space nor the primordial gaussian adiabatic fluctuations derive directly from general relativity and the standard model. These can be tested, again one by one, by extensions of the parameter set which concerns the early universe physics. In the baseline ΛCDM model, there are only two parameters describing

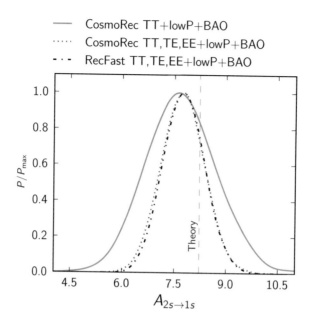

FIGURE 8. The marginalized posterior distributions are shown for A_{2s-1s} with different theoretical CMB codes and data combinations. The theoretical value is shown by the dashed line. (Ref. [21].)

the initial conditions. One is the amplitude of the initial fluctuations A_s for which no generic prediction exists. The other one is the spectral index n_s of the power law spectrum describing the primordial fluctuations. The inflation paradigm of an exponential acceleration of the expansion associated with quantum fluctuations of the inflation field gives a natural common origin to these assumptions and explains the flat space geometry. Quantum fluctuations as the sources of the primordial energy density fluctuations have been proposed by Mukhanov and Chibisov [25]. This model is very appealing because quantum fluctuations should arise in the early universe and no other source has been proposed. In this paradigm, inflation brings the quantum fluctuations to macroscopic scales and ends with a reheating phase where the particles and fields we know are generated. It leaves adiabatic gaussian fluctuations of energy density.

Testing the Gaussianity uses three point correlations after removing the expected part from the two point correlations and also the ISW and lensing contributions. The values for the different geometries of the three points are given in Table 3. The adiabaticity of the fluctuations can be tested by putting an upper limit on the isocurvature fluctuations which had been one model explored

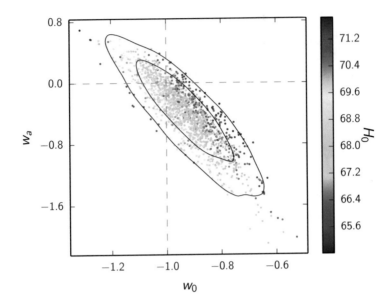

FIGURE 9. The sample distribution in the $w_0 - w_a$ plane of the dark energy parameters are shown color coded as a function of H_0. The contour are the 68% and 95% limits. Dashed gray lines intersect at the point corresponding to a cosmological constant. (Ref. [21].)

before ΛCDM. The fraction of isocurvature modes can be described by a parameter $\alpha = 0.0003 \pm 0.0016$. These are only upper limits and are very low, in full agreement with the inflation prediction. They are already getting close to values dominated by other effects and thus even if sensitivity improves would probably not give much clues on the primordial universe physics.

There are two more interesting parameters to constrain early universe physics. The end of inflation is not sudden implying that large and small scales do not reach the horizon at the same time. This implies that the anisotropies power spectrum index n_s should deviate from the Harrison-Zeldovich value 1 and be slightly smaller. The deviation from 1 is predicted to be of order $1/N$ where N ($\sim$ 50 – 60) leading to $n_s \sim 1 - 1/N = 0.96 - 0.97$ (N is the number of e-foldings between the end of inflation and the time our present day Hubble scale crossed the inflation horizon). This deviation from scale invariance is shown by *Planck* to be present with a $6\,\sigma$ confidence level in the ΛCDM parameters (Table 2). This is one of the major results from Planck:

$$n_{\rm s} = 0.9667 \pm 0.0040.$$

Shape and method	f_{NL}(KSW)	
	Independent	ISW-lensing subtracted
SMICA (T)		
Local	10.2 ± 5.7	2.5 ± 5.7
Equilateral	-13 ± 70	-16 ± 70
Orthogonal	-56 ± 33	-34 ± 33
SMICA ($T + E$)		
Local	6.5 ± 5.0	0.8 ± 5.0
Equilateral	3 ± 43	-4 ± 43
Orthogonal	-36 ± 21	-26 ± 21

TABLE 3. Limits of the f_{NL} non-Gaussianity parameter from temperature only (top) and combined with Polarization (bottom). The right column shows the results after subtraction of the ISW and lensing effects. This leaves upper limits on the primordial non-Gaussianity. (Ref. [24].)

It is possible to introduce other parameters related to primordial physics in the *Planck* analysis. The inflation models also predicts the flat space because the exponential expansion flattens the space geometry such that the curvature would not be detectable today. The best value for the residual contribution of curvature to the total energy density is

$$\Omega_K = 0.000 \pm 0.005 \; (95\% \text{ confidence level}).$$

A potential curvature of the primordial spectrum is an extension which allows to explore more complicated models. *Planck* finds a low upper limit:

$$dn_s/d\ln k = 0.0065 \pm 0.0076.$$

Inflation models generate a fraction r of the energy density fluctuations in form of tensor fluctuations (gravity waves) which would imprint B-modes polarization on the CMB although some models predicts values of the r parameter much lower than we can detect. Figure 10 shows the constraints from *Planck* for the allowed regions in the $r - n_s$ plane together with the predictions of models and classes of inflation models.

An early claim by the Bicep2 team for detection of the primordial B-modes assumed wrongly a negligible dust contribution. Nevertheless, the *Planck* data showed evidence for an interstellar dust contribution close to the observed level of the Bicep2 measurement. Currently, the tightest B-modes constraint on r comes from a common Bicep2-Keck-*Planck* analysis [22] which gives

$$r_{0.002} < 0.08 \text{ with } 95\% \text{ confidence level}.$$

This makes the standard particles physics and cosmology models an ensemble which accounts for many observations and experimental results. They also both

J.-L. Puget

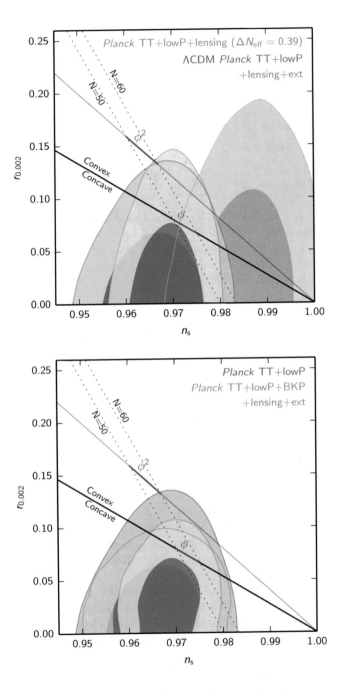

FIGURE 10. Plots of non excluded regions in the plane $n_s - r$. Positions of classes of inflation models are also shown. The grey area shows the feet of relaxing the effective number of neutrinos species. (Ref. [21].)

need to be extended coherently with new physics to resolve common questions (nature of dark energy and dark matter, supersymmetry, physics near the Planck scale...). *Planck* data is a major step in reinforcing the inflation paradigm of the early universe and requiring new physics but has only been able to remove large classes of inflation models.

References

[1] Alpher, R. A., Bethe, H., & Gamow, G.: Phys. Rev. **73**, 803 (1948).

[2] Penzias, A. A., & Wilson, R. W.: ApJ **142**, 419 (1965).

[3] Dicke, R. H., Peebles, P. J. E., Roll, P. G., & Wilkinson, D. T.: ApJ **142**, 414 (1965).

[4] Zwicky, F.: Helvetica Physica Acta **6**, 110 (1933).

[5] Zwicky, F.: ApJ **86**, 217 (1937).

[6] Mather, J. C., Fixsen, D. J., Shafer, R. A., Mosier, C., & Wilkinson, D. T.: ApJ **512**, 511 (1999).

[7] Zeldovich, Y. B., Kurt, V. G., & Syunyaev, R. A.: Zhurnal Eksperimentalnoi i Teoreticheskoi Fiziki **55**, 278 (1968).

[8] Peebles, P. J. E., & Yu, J. T.: ApJ **162**, 815 (1970).

[9] Harrison, E. R.: PRD **1**, 2726 (1970).

[10] Zeldovich, Y. B.: MNRAS **160**, 1P (1972).

[11] Smoot, G. F., Bennett, C. L., Kogut, A., et al.: ApJ **396**, L1 (1992).

[12] Peebles, P. J. E.: ApJ **483**, L1 (1997).

[13] Riess, A. G., Filippenko, A. V., Challis, P., et al.: AJ **116**, 1009 (1998).

[14] Perlmutter, S., Turner, M. S., & White, M.: Phys. Rev. Lett. **83**, 670 (1999).

[15] Sakharov, A. D.: Soviet Journal of Experimental and Theoretical Physics **22**, 241 (1966).

[16] Planck Collaboration, Ade, P. A. R., Aghanim, N., et al.: A&A **536**, A2 (2011).

[17] de Bernardis, P., Ade, P. A. R., Bock, J. J., et al.: Nat. **404**, 955 (2000).

[18] Hanany, S., Ade, P., Balbi, A., et al.: ApJl **545**, L5 (2000).

[19] Spergel, D. N., Verde, L., Peiris, H. V., et al.: ApJs **148**, 175 (2003).

[20] Planck Collaboration, Adam, R., Ade, P. A. R., et al.: AAP **594**, A1 (2016).

[21] Planck Collaboration, Ade, P. A. R., Aghanim, N., et al.: AAP **594**, A13 (2016).

[22] BICEP2/Keck Collaboration, Planck Collaboration, Ade, P. A. R., et al.: Phys. Rev. Lett. **114**, 101301 (2015).

[23] Planck Collaboration, Aghanim, N., Ashdown, M., et al.: A&A **596**, A107 (2016).

[24] Planck Collaboration, Ade, P. A. R., Aghanim, N., et al.: AAP **594**, A17 (2016).

[25] Mukhanov, V. F., & Chibisov, G. V.: Soviet Journal of Experimental and Theoretical Physics Letters **33**, 532 (1981).

Jean-Loup Puget
IAS
Bât. 121
Université Paris Sud
91405 Orsay cedex
France
e-mail: jean-loup.puget@ias.u-psud.fr

The Universe, 93–119

Massive Black Holes: Evidence, Demographics and Cosmic Evolution

Reinhard Genzel

Abstract. The article summarizes the observational evidence for the existence of massive black holes, as well as the current knowledge about their abundance, their mass and spin distributions, and their cosmic evolution within and together with their galactic hosts. We finish with a discussion of how massive black holes may in the future serve as laboratories for testing the theory of gravitation in the extreme curvature regimes near the event horizon.

Keywords. Massive black holes, galactic center black hole, orbiting stars, local universe, cosmic evolution.

Rapporteur talk 'Black holes', in: Astrophysics and Cosmology, Proceedings of the 26th Solvay Conference on Physics, R. Blandford, D. Gross and A. Sevrin, eds., ©2016 World Scientific

1. Introduction

In 1784 Rev. John Michell was the first to note that a sufficiently compact star may have a surface escape velocity exceeding the speed of light. He argued that an object of the mass of the Sun (or larger) but with a radius of 3 km (instead of the Sun's radius of 700,000 km) would thus be invisible. A proper mathematical treatment of this problem then had to await Albert Einstein's General Relativity ("GR", 1916). Karl Schwarzschild's (1916) solution of the vacuum field equations in spherical symmetry demonstrated the existence of a characteristic event horizon, the Schwarzschild radius $R_s = 2GM/c^2$, within which no communication is possible with external observers. Roy Kerr (1963) generalized this solution to spinning black holes. The mathematical concept of a black hole was established (although the term itself was coined only later by John Wheeler in 1968). In GR, all matter within the event horizon is predicted to be inexorably pulled toward the center where all gravitational energy density (matter) is located in a density singularity. From considerations of the information content of black holes, there is significant tension between the predictions of GR and Quantum theory (e.g. Susskind 1995, Maldacena 1998, Bousso 2002). It is generally thought that a proper quantum theory of gravity will modify the concepts of GR on scales comparable to or smaller than the Planck length, $l_{Pl} \sim 1.6 \times 10^{-33}$cm, remove the concept of a central singularity, and potentially challenge the interpretation of the GR event horizon (Almheiri et al. 2013).

But are these objects of GR realized in Nature?

2. First Evidence

Astronomical evidence for the existence of black holes started to emerge in the 1960s with the discovery of distant luminous 'quasi-stellar-radio-sources/objects' (QSOs, Schmidt 1963) and variable X-ray emitting binaries in the Milky Way (Giacconi et al. 1962). It became clear from simple energetic arguments that the enormous luminosities and energy densities of QSOs (up to several 10^{14} times the luminosity of the Sun, and several 10^4 times the entire energy output of the Milky Way Galaxy), as well as their strong UV-, X-ray and radio emission can most plausibly be explained by accretion of matter onto massive black holes (e.g. Lynden-Bell 1969, Shakura & Sunyaev 1973, Rees 1984, Blandford 1999). Simple theoretical considerations show that between 7% (for a non-rotating Schwarzschild hole) and 40% (for a maximally rotating Kerr hole) of the rest energy of an infalling particle can in principle be converted to radiation outside the event horizon, a factor 10 to 100 more than in stellar fusion from hydrogen to helium. To explain powerful quasars by this mechanism, black hole masses of 10^8 to 10^9 solar masses and accretion flows between 0.1 to 10 solar masses per year are required. QSOs are located (without exception) at the nuclei of large, massive galaxies (e.g. Osmer 2004). QSOs just represent the most extreme and spectacular among the general nuclear activity of most galaxies. This includes variable X- and γ-ray emission and highly collimated, relativistic radio jets, all of which cannot be accounted for by stellar activity.

The 1960s and 1970s brought also the discovery of X-ray stellar binary systems (see Giacconi 2003 for an historic account). For about 20 of these compact and highly variable X-ray sources dynamical mass determinations from Doppler spectroscopy of the visible primary star established that the mass of the X-ray emitting secondary is significantly larger than the maximum stable neutron star mass, ~ 3 solar masses (McClintock & Remillard 2004, Remillard & McClintock 2006, Özel et al. 2010). The binary X-ray sources thus are excellent candidates for stellar black holes (SBH). They are probably formed when a massive star explodes as a supernova at the end of its fusion lifetime and the compact remnant collapses to a stellar hole.

An unambiguous proof of the existence of a stellar or massive black hole, as defined by GR, requires the determination of the gravitational potential to the scale of the event horizon. This proof can in principle be obtained from spatially resolved measurements of the motions of test particles (interstellar gas or stars) in close orbit around the black hole. In practice it is not possible (yet) to probe the scale of an event horizon of any black hole candidate (SBH as well as MBH) with spatially resolved dynamical measurements. A more modest goal then is to show that the gravitational potential of a galaxy nucleus is dominated by a compact non-stellar mass and that this central mass concentration cannot be anything but a black hole because all other conceivable configurations are more extended, are not stable, or produce more light (e.g. Maoz 1995, 1998). Even this test cannot be conducted yet in distant QSOs from dynamical measurements. It has become feasible over the last decades in nearby galaxy nuclei, however, including the Center of our Milky Way.

3. NGC 4258

Solid evidence for central 'dark' (i.e., non-stellar) mass concentrations in about 80 nearby galaxies has emerged over the past two decades (e.g. Magorrian 1998, Kormendy 2004, Gültekin et al. 2009, Kormendy & Ho 2013, McConnell & Ma 2013) from optical/infrared imaging and spectroscopy on the Hubble Space Telescope (HST) and large ground-based telescopes, as well as from Very Long Baseline radio Interferometry (VLBI).

The first truly compelling case that such a dark mass concentration cannot just be a dense nuclear cluster of white dwarfs, neutron stars and perhaps stellar black holes emerged in the mid-1990s from spectacular VLBI observations of the nucleus of NGC 4258, a mildly active galaxy at a distance of 7 Mpc (Miyoshi et al. 1995, Moran 2008, Figure 1). The VLBI observations show that the galaxy nucleus contains a thin, slightly warped disk of H_2O masers (viewed almost edge on) in Keplerian rotation around an unresolved mass of 40 million solar masses (Figure 1). The inferred density of this mass exceeds a few 10^9 solar masses pc^{-3} and thus cannot be a long-lived cluster of 'dark' astrophysical objects of the type mentioned above (Maoz 1995). As we will discuss below, a still more compelling case can be made in the case of the Galactic Center.

4. The Galactic Center Black Hole

The central light years of our Galaxy contain a dense and luminous star cluster, as well as several components of neutral, ionized and extremely hot gas (Genzel, Hollenbach & Townes 1994, Genzel, Eisenhauer & Gillessen 2010). The central dark mass concentration discussed above is associated with the compact radio source SgrA*, which has a size of about 10 light minutes and is located at the center of the nuclear star cluster. SgrA* thus may be a MBH analogous to QSOs, albeit with orders of magnitude lower mass and luminosity. Because of its proximity - the distance to the Galactic Center is about 8.3 kilo-parsecs (kpc), about 10^5 time closer than the nearest QSOs - high resolution observations of the Milky Way

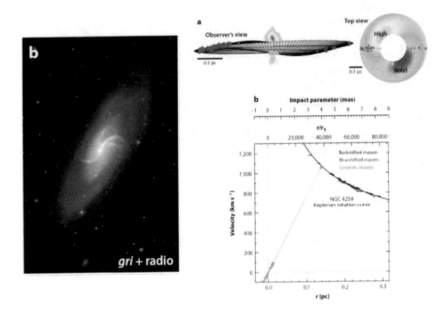

FIGURE 1. Left: Optical and radio image of the active galaxy
NGC4258. This disk galaxy exhibits a spectacular curved twin
radio and X-ray jet, visible in orange in this picture. Right: (top)
Schematic edge-on (left) and face-on (right) views of the almost-
edge-on, warped maser disk of NGC 4258 (from Moran 2008) with
warp parameters from Herrnstein et al. (2005) and including the
inner contours of the radio jet. The relative positions of the re-
ceding, near-systemic, and approaching H_2O masers are indicated
by red, green, and blue spots, respectively. Differences in line-of-
sight projection corrections to the slightly tilted maser velocities
account for the departures in the high-velocity masers from exact
Keplerian rotation. The near-systemic masers are seen tangent to
the bottom of the maser disk bowl along the line of sight. They
drift from right to left in ~12 years across the green areas where
amplification of the background radio continuum is sufficient for
detection. (b) NGC 4258 rotation velocity versus radius in units
of parsec (bottom axis), Schwarzschild radii (top axis), and mil-
liarcsec (extra axis). The black curve is a Keplerian fit to 4255
velocities of red- and blue-shifted masers (red and blue dots). The
small green points and line show 10036 velocities of near-systemic
masers and a linear fit to them. The green filled circle is the corresponding mean velocity point. The maser data are taken from
Argon et al. (2007) (adapted from Kormendy & Ho 2013).

nucleus offer the unique opportunity of carrying out a stringent test of the MBH-paradigm and of studying stars and gas in the immediate vicinity of a MBH, at a level of detail that will not be accessible in any other galactic nucleus for the foreseeable future. Since the Center of the Milky Way is highly obscured by interstellar dust particles in the plane of the Galactic disk, observations in the visible part of the electromagnetic spectrum are not possible. The veil of dust, however, becomes transparent at longer wavelengths (the infrared, microwave and radio bands), as well as at shorter wavelengths (hard X-ray and γ-ray bands), where observations of the Galactic Center thus become feasible.

The key obviously lies in very high angular resolution observations. The Schwarzschild radius of a 4 million solar mass black hole at the Galactic Center subtends a mere 10^{-5} arc-seconds[1]. For high resolution imaging from the ground an important technical hurdle is the correction of the distortions of an incoming electromagnetic wave by the refractive Earth atmosphere. For some time radio astronomers have been able to achieve sub-milli-arcsecond resolution VLBI at millimeter wavelengths, with the help of phase-referencing to nearby compact radio sources. In the optical/near-infrared waveband the atmosphere distorts the incoming electromagnetic waves on time scales of milliseconds and smears out long-exposure images to a diameter of more than an order of magnitude greater than the diffraction limited resolution of large ground-based telescopes (Figure 2). From the early 1990s onward initially 'speckle imaging' (recording short exposure images, which are subsequently processed and co-added to retrieve the diffraction limited resolution and then later 'adaptive optics' (AO: correcting the wave distortions on-line) became available, which have since allowed increasingly precise high resolution near-infrared observations with the currently largest (10 m diameter) ground-based telescopes of the Galactic Center (and nearby galaxy nuclei).

Early evidence for the presence of a non-stellar mass concentration of 2-4 million times the mass of the Sun ($M_{\odot}$) came from mid-infrared imaging spectroscopy of the 12.8 μm [NeII] line, which traces emission from ionized gas clouds in the central parsec region (Wollman et al. 1977, Lacy et al. 1980, Serabyn & Lacy 1985). However, many considered this dynamical evidence not compelling because of the possibility of the ionized gas being affected by non-gravitational forces (shocks, winds, magnetic fields). A far better probe of the gravitational field are stellar motions, which started to become available from Doppler spectroscopy in the late 1980s. They confirmed the gas motions (Rieke & Rieke 1988, McGinn et al. 1989, Sellgren et al. 1990, Krabbe et al. 1995, Haller et al. 1996, Genzel et al. 1996). The ultimate breakthrough came from the combination of AO techniques with advanced imaging and spectroscopic instruments (e.g. 'integral field' imaging spectroscopy, Eisenhauer et al. 2005) that allowed diffraction limited near-infrared spectroscopy and imaging astrometry with a precision initially at the few milli-arcsecond scale, and improving to a few hundred micro-arcseconds in the next decade (c.f. Ghez et al. 2008, Gillessen et al. 2009). With diffraction

---

[1]10 μarc-seconds correspond to about 2 cm at the distance of the Moon.

R. Genzel

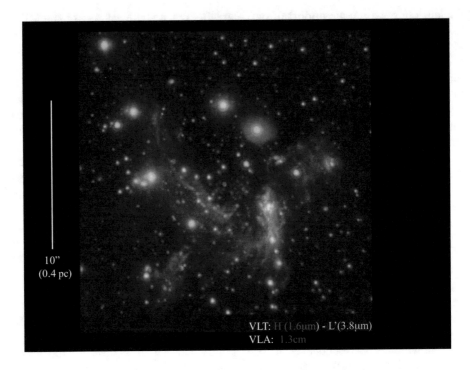

FIGURE 2. Near-infrared/radio, color-composite image of the central light years of Galactic Center. The blue and green colors represent the 1.6 and 3.8μm broad band near-infrared emission, at the diffraction limit (∼0.05") of the 8m Very Large Telescope (VLT) of the European Southern Observatory (ESO), and taken with the 'NACO' AO-camera and an infrared wavefront sensor (adapted from Genzel et al. 2003). Similar work has been carried out at the 10 m Keck telescope (Ghez et al. 2003, 2005). The red color image is the 1.3 cm radio continuum emission taken with the Very Large Array (VLA) of the US National Radio Astronomy Observatory (NRAO). The compact red dot in the center of the image is the compact, non-thermal radio source SgrA*. Many of the bright blue stars are young, massive O/B- and Wolf-Rayet stars that have formed recently. Other bright stars are old, giants and asymptotic giant branch stars in the old nuclear star cluster. The extended streamers/wisps of 3.8 μm emission and radio emission are dusty filaments of ionized gas orbiting in the central light years (adapted from Genzel, Eisenhauer & Gillessen 2010).

limited imagery starting in 1992 on the 3.5m New Technology Telescope (NTT) of the European Southern Observatory (ESO) in La Silla/Chile, and continuing since 2002 on ESO's Very Large Telescope (VLT) on Paranal, a group at MPE was able to determine proper motions of stars as close as $\sim$0.1" from SgrA* (Eckart & Genzel 1996, 1997). In 1995 a group at the University of California, Los Angeles started a similar program with the 10m diameter Keck telescope in Hawaii (Ghez et al. 1998). Both groups independently found that the stellar velocities follow a 'Kepler' law ($v \sim R^{-1/2}$) as a function of distance from SgrA* and reach $\geq 10^3$ km/s within the central light month.

Only a few years later both groups achieved the next and crucial steps. Ghez et al. (2000) detected accelerations for three of the 'S'-stars, Schödel et al. (2002) and Ghez et al. (2003) showed that the star S2/S02 is in a highly elliptical orbit around the position of the radio source SgrA*, and Schödel et al. (2003) and Ghez et al. (2005) determined the orbits of 6 additional stars. In addition to the proper motion/astrometric studies, they obtained diffraction limited Doppler spectroscopy of the same stars (Ghez et al. 2003, Eisenhauer et al. 2003, 2005), allowing precision measurement of the three dimensional structure of the orbits, as well as the distance to the Galactic Center. Figure 3 shows the data and best fitting Kepler orbit for S2/S02, the most spectacular of these stars with a 16 year orbital period (Ghez et al. 2008, Gillessen et al. 2009, 2009a). At the time of writing, the two groups have determined individual orbits for more than 40 stars in the central light month. These orbits show that the gravitational potential indeed is that of a point mass centered on SgrA*. These stars orbit the position of the radio source SgrA* like planets around the Sun. The point mass must be concentrated well within the peri-approaches of the innermost stars, $\sim$10-17 light hours, or 70 times the Earth orbit radius and about 1000 times the event horizon of a 4 million solar mass black hole. There is presently no indication for an extended mass greater than about 2 % of the point mass.

VLBI observations have set an upper limit of about 20 km/s and 2 km/s to the motion of SgrA* itself, along and perpendicular to the plane of the Milky Way, respectively (Reid & Brunthaler 2004). When compared to the two orders of magnitude greater velocities of the stars in the immediate vicinity of SgrA*, this demonstrates that the radio source must indeed be massive, with simulations giving a lower limit to the mass of SgrA* of $\sim 10^5$ solar masses (Chatterjee, Hernquist & Loeb 2002). The intrinsic size of the radio source at about 1mm is only about 4 times the event horizon diameter of a 4 million solar mass black hole (Bower et al. 2004, Shen et al. 2005, Doeleman et al. 2008). Combining radio size and proper motion limit of SgrA* with the dynamical measurements of the nearby orbiting stars leads to the conclusion that SgrA* can only be a massive black hole, beyond any reasonable doubt (Genzel et al. 2010).

The current Galactic Center evidence eliminates all plausible astrophysical plausible alternatives to a massive black hole. These include astrophysical clusters of neutron stars, stellar black holes, brown dwarfs and stellar remnants (e.g., Maoz

R. Genzel

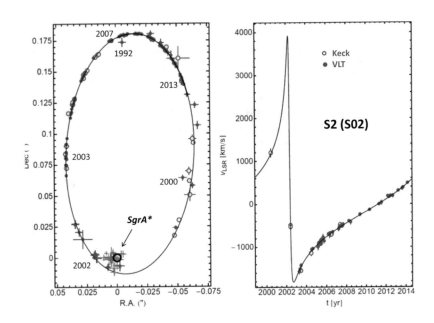

FIGURE 3. Position on the sky as a function of time (left) and
Doppler velocity (relative to the Local Standard of Rest) as a
function of time (right) of the star 'S2 (or S02)' orbiting the compact radio source SgrA*. Blue filled circles denote data taken with
the ESO NTT and VLT as part of the MPE Galactic Center monitoring project (Schödel et al. 2002, 2005, Gillessen et al. 2009),
and red open circles denote data taken with the Keck telescope as
part of the UCLA monitoring project (Ghez et al. 2003, 2008, see
Gillessen et al. 2009a for the calibration to a common reference
frame). Superposed is the best fitting elliptical orbit (continuous
curve: central mass 4.26 (± 0.14)$_{statistical}$ (± 0.2)$_{systematics}$ million
solar masses, distance 8.36 (± 0.1)$_{stat}$ (± 0.15)$_{syst}$ kpc) with its
focus at (0,0) in the left inset (including the independent distance constraints of Reid et al. 2014, Chatzopoulos et al. 2014).
The astrometric position of SgrA* is denoted by a circle, grey
crosses mark the locations of infrared flares (of typical duration
1-3 hours) that are believed to originate from within the immediate vicinity of the event horizon. The radio source is coincident
within the 2 milli-arcsecond errors with the gravitational centroid
of the stellar orbit. Since the beginning of the MPE monitoring
project (1991/1992), the star has completed its first full orbit in
2007, and it passed its peri-center position 17 light hours from
SgrA* in spring 2002 (and again in spring 2018).

1995, 1998; Genzel et al. 1997, 2000; Ghez et al. 1998, 2005), and even fermion balls (Viollier, Trautmann & Tupper 1993, Munyaneza, Tsiklauri & Viollier 1998, Ghez et al. 2005; Genzel, Eisenhauer & Gillessen 2010). Clusters of a very large number of mini-black holes and boson balls (Torres, Capozziello & Lambiase 2000; Schunck & Mielke 2003; Liebling & Palenzuela 2012) are harder to exclude. The former have a large relaxation and collapse time, the latter have no hard surfaces that could exclude them from luminosity arguments (Broderick, Loeb & Narayan 2009), and they are consistent with the dynamical mass and size constraints. However, such a boson 'star' would be unstable to collapse to a MBH when continuously accreting baryons (as in the Galactic Center), and it is very unclear how it could have formed. Under the assumption of the validity of General Relativity the Galactic Center is now the best quantitative evidence that MBH do indeed exist.

5. Massive Black Holes in the local Universe

Beyond the "gold standards" in the Galactic Center and NGC 4258, evidence for the presence of central mass concentrations (which we will henceforth assume to be MBH even though this conclusion can be challenged in most of the individual cases), and a census of their abundance and mass spectrum comes from a number of independent methods,

- robust evidence for MBH in about 10 galaxies comes from VLBI studies of H_2O maser spots in circum-nuclear Keplerian disks of megamaser galaxies akin to NGC4258 (the NRAO "megamaser cosmology project", `https://safe.nrao.edu/wiki/bin/view/Main/MegamaserCosmologyProject`, Braatz et al. 2010, Kuo et al. 2011, Reid et al. 2013);
- robust evidence for MBH for about 80 galaxies comes from modeling of the spatially resolved, line-of-sight integrated stellar Doppler-velocity distributions with the Hubble Space Telescope (HST) and large ground based telescopes with AO (see the recent reviews of Kormendy & Ho 2013, McConnell & Ma 2013 and references therein). Among the latter, a particularly impressive case is the nucleus of M31, the Andromeda galaxy, where a 10^8 $M_\odot$ central mass is identified from the rapid ($\sim$900 km/s) rotation of a compact circum-nuclear stellar disk (Bender et al. 2005);
- for a number of galaxies, observations of the spatially resolved motions of ionized gas also provide valuable evidence for central mass concentrations, which, however, can be challenged, as mentioned for the Galactic Center, by the possibility of non-gravitational motions (Macchetto et al. 1997, van der Marel & van den Bosch 1998, Barth et al. 2001, Marconi et al. 2003, 2006, Neumayer et al. 2006);
- most recently, high resolution interferometric observations of CO emission have become available as a promising new tool for determining robust central masses (Davis et al. 2013);

- qualitative evidence for the presence of accreting black holes naturally comes for all bona-fide AGN from their IR-optical-UV- and X-ray spectral signatures. In the case of type 1 AGN with broad permitted lines coming from the central light days to light years around the black hole (c.f. Netzer 2013 and references therein), it is possible to derive the size of the broad line region (BLR) from correlating the time variability of the (extended) BLR line emission with that of the (compact) ionizing UV continuum. This reverberation technique (Blandford & McKee 1982) has been successfully applied to derive the BLR sizes (Peterson 1993, 2003, Netzer & Peterson 1997, Kaspi et al. 2000) in several dozen AGN, and has yielded spatially resolved imaging of the BLR in a few (e.g. Bentz et al. 2011). Kaspi et al. (2000). These observations show that the size of the BLR is correlated with the AGN optical luminosity, $R_{BLR} \sim [(\nu L_\nu)_{5100 \text{ Å}}]^{0.7}$. After empirical calibration of the zero points of the correlation measurements of the line width of the BLR and of the rest frame optical luminosity of the AGN are sufficient to make an estimate of the MBH mass. As this requires only spectro-photometric data, the technique can be applied even for distant (high redshift) type 1 AGNs (Vestergaard 2004, Netzer et al. 2006, Traktenbrot & Netzer 2012), as well as for low-luminosity AGN in late type and dwarf galaxies (Filippenko & Sargent 1989, Ho, Filippenko & Sargent 1997, Greene & Ho 2004, 2007, Ho 2008, Reines et al. 2011, Greene 2012, Reines, Greene & Geha 2013).

6. Demographics and MBH-galaxy "co-evolution"

These data give a fairly detailed census of the incidence and of the mass spectrum of the local (and less so, also of the distant) MBH population. MBH masses span a range at least five orders of magnitudes from 10^5 M$_\odot$ in dwarf galaxies to 10^{10} M$_\odot$ in the most massive central cluster galaxies. Most massive spheroidal/bulged galaxies appear to have a central MBH. The occupation fraction drops in bulgeless systems with decreasing galaxy mass (Greene 2012). It is not clear yet whether the lack of observational evidence below 10^5 M$_\odot$ is real, or driven by observational detectability. The inferred black hole mass and the mass of the galaxy's spheroidal component (but not its disk, or dark matter halo) are strongly correlated (Magorrian et al. 1998, Häring & Rix 2004). The most recent analyses of Kormendy & Ho (2013) and McConnell & Ma (2013) find that between 0.3 and 0.5% of the bulge/spheroid mass is in the central MBH. The scatter of this relation is between ±0.3 and ±0.5 dex, depending on sample and analysis method (McConnell & Ma 2013). A correlation of comparable scatter exists between the black hole mass and the bulge/spheroid velocity dispersion σ (M$_{BH} \sim \sigma^\beta$, with $\beta \sim 4.2 - 5.5$, Ferrarese & Merritt 2000, Gebhardt et al. 2000, Tremaine et al. 2002, Kormendy & Ho 2013, McConnell & Ma 2013, Figure 4).

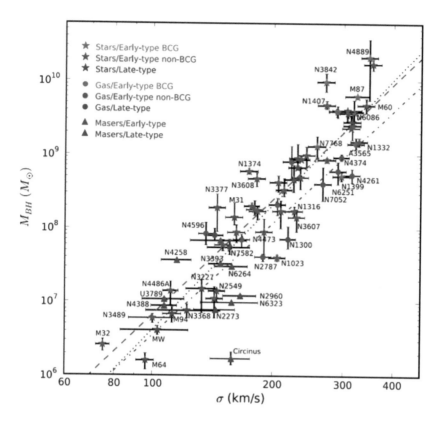

FIGURE 4. Black hole mass MBH (vertical axis) as a function of galaxy velocity dispersion σ (horizontal axis), for the all 72 galaxies in the compendium of McConnell & Ma (2013). Asterisks, filled circles and filled triangles denote the technique that was used to determine the MBH mass (stellar kinematics, gas kinematics, or masers), and red, green and blue colors denote the type of host galaxy (spheroidal galaxy, very massive spheroidal galaxy at the center of a galaxy cluster (BCG), and late type (disk/ irregular) galaxy). The black dotted line shows the best-fitting power law for the entire sample: $\log 10(MB_H/M_\odot) = 8.32 + 5.64 \log(\sigma/200\mathrm{km/s})$. When early-type and late-type galaxies are fitted separately, the resulting power laws are $\log(MB_H/M_\odot) = 8.39 + 5.20 \log(\sigma/\mathrm{km/s})$ for the early-type (red dashed line), and $\log(MB_H/M_\odot) = 8.07 + 5.06 \log(\sigma/200\mathrm{km/s})$ for the late-type galaxies (blue dot-dashed line). The plotted values of σ are derived using kinematic data within the effective radius of the spheroidal galaxy component (adapted from McConnell & Ma 2013).

Ever since this correlation between central black hole mass and galaxy host spheroidal mass (or velocity dispersion) component has been established, the interpretation has been that there must be an underlying connection between the formation paths of the galaxies' stellar components and their embedded central MBHs. This underlying connection points back to the peak formation epoch of massive galaxies about 6-10 Gyrs ago (e.g. Madau et al. 1996, Haehnelt 2004). The fact that the correlation is between the black hole mass and the bulge/spheroidal component, and not the total galaxy or dark matter mass, has been taken as evidence that most of the MBH's growth, following an early evolution from a lower mass seed, is triggered by a violent dissipative process at this early epoch. The most obvious candidate are major mergers between early gas rich galaxies, which are widely thought to form bulges in the process (Barnes & Hernquist 1996, Kauffmann & Haehnelt 2000, Haiman & Quataert 2004, Hopkins et al. 2006, Heckman et al. 2004). Compelling support for the AGN – merger model comes from the empirical evidence that dusty ultra-luminous infrared galaxies (ULIRGs, $L_{IR} > 10^{12}$ $L_{\odot}$) in the local Universe are invariably major mergers of gas-rich disk galaxies (Sanders et al. 1988); the majority of the most luminous late stage ULIRGs are powered by obscured AGN (Veilleux et al. 1999, 2009).

This 'strong' co-evolution model is further supported by the fact that the peak of cosmic star formation 10 Gyrs ago is approximately coeval with the peak of cosmic QSO activity (Boyle et al. 2000), and that the amount of radiation produced during this QSO era is consistent with the mass present in MBHs locally for a 10-20% radiation efficiency during MBH mass growth (Soltan 1982, Yu et al. 2002, Marconi et al. 2004, Shankar et al. 2009). There is an intense ongoing discussion whether or not MBHs and their hosts galaxies formed coevally and grew on average in lock-step (Figure 5, Marconi et al. 2004, Shankar et al. 2009, Alexander & Hickox 2012, Mullaney et al. 2012, del Vecchio et al. 2014), or whether MBHs started slightly earlier or grew more efficiently (Jahnke et al. 2009, Merloni et al. 2010, Bennett et al. 2011). The fact that the correlation appears to be quite tight suggests that feedback between the accreting and rapidly growing black holes during that era and the host galaxy may have been an important contributor to the universal shutdown of star formation and mass growth in galaxies above the Schechter mass, $M_S \geq 10^{10.9}$ $M_{\odot}$ (Baldry et al. 2008, Conroy & Wechsler 2009, Peng et al. 2010, Moster et al. 2013, Behroozi et al. 2013).

7. AGN-MBH feedback

Throughout the last 10 billion years galaxies have been fairly inefficient in incorporating the cosmic baryons available to them into their stellar components. At a dark matter halo mass near 10^{12} $M_{\odot}$ this baryon fraction is only about 20% (of the cosmic baryon abundance), and the efficiency drops to even lower values on either side of this mass (e.g., Madau et al. 1996; Baldry et al. 2008, Conroy & Wechsler 2009, Guo et al. 2010, Moster et al. 2013, Behroozi et al. 2013). Galactic

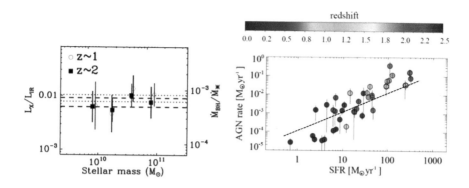

FIGURE 5. Evidence for average MBH-galaxy growth co-evolution by stacking deep X-ray data (as quantitative indicators of AGN growth) on multi-wavelength images of star forming galaxies (including mid- and far-IR emission as extinction- and (nearly) AGN-independent tracers of star formation rates) in GOODS-S (left, Mullaney et al. 2012) and GOODS-S and COSMOS (right, delVecchio et al. 2014). The left plot shows that the ratio of inferred black hole growth to star formation rate is the same in several mass bins and at $z \sim 1$ and $z \sim 2$. The right plot shows the integrated MBH growth rate as a function of star formation and redshift (colors). The dotted line has slope unity and is not a fit to the data. However, once the dependence on mass and redshift are de-coupled, the best fitting correlation does have unity slope, suggesting average co-evolution.

winds driven by supernovae and massive stars have long been proposed to explain the low baryon content of halos much below $\log(M_h/M_\odot) \sim 12$ (e.g. Dekel & Silk 1986, Efstathiou 2000). The decreasing efficiency of galaxy formation above $\log(M_h/M_\odot) \sim 12$ may be caused by less efficient cooling and accretion of baryons in massive halos (Rees & Ostriker 1977, Dekel & Birnboim 2006). Alternatively or additionally efficient outflows driven by accreting MBH may quench star formation at the high mass tail, at and above the Schechter stellar mass, $M_S \sim 10^{10.9}\ M_\odot$ (di Matteo, Springel & Hernquist 2005, Croton et al. 2006, Bower et al. 2006, Hopkins et al. 2006, Cattaneo et al. 2007, Somerville et al. 2008, Fabian 2012).

In the local Universe, such 'AGN or MBH feedback' has been observed in the so called 'radio mode' in very massive, central cluster galaxies driving jets into the intra-cluster medium. In these cases the central MBHs typically is in a fairly low or quiescent radiative state. Considerations of energetics suggest that radio mode feedback plausibly prevents cooling cluster gas to fall onto these massive galaxies that would otherwise lead to substantial further star formation and mass growth (McNamara & Nulsen 2007, Fabian 2012, Heckman & Best 2014). A second MBH feedback mode (termed 'QSO mode'), in which the MBH is active (i.e., the AGN is luminous) is detected as ionized winds from AGN (e.g. Cecil, Bland, & Tully 1990, Veilleux, Cecil & Bland-Hawthorn 2005, Westmoquette et al. 2012, Rupke & Veilleux 2013, Harrison et al. 2014) and from obscured QSOs (Zakamska & Greene 2014). The QSO mode feedback in form of powerful neutral and ionized gas outflows has also been found in late stage, gas rich mergers (Fischer et al. 2010, Feruglio et al. 2010, Sturm et al. 2011, Rupke & Veilleux 2013, Veilleux et al. 2013), which however are rare in the local Universe.

At high-z AGN QSO mode feedback has been seen in broad absorption line quasars (Arav et al. 2001, 2008, 2013, Korista et al. 2008), in type 2 AGN (Alexander et al. 2010, Nesvadba et al. 2011, Cano Díaz et al. 2012, Harrison et al. 2012), and in radio galaxies (Nesvadba et al. 2008). However, luminous AGNs near the Eddington limit are again rare, constituting less than 1% of the star forming population in the same mass range (e.g. Boyle et al. 2000). QSOs have short lifetimes relative to the Hubble time ($t_{QSO} \sim 10^7 - 10^8$ yr $\ll t_H$, Martini 2004) and thus have low duty cycles compared to galactic star formation processes ($t_{SF} \sim 10^9$ yr, Hickox et al. 2014). It is thus not clear whether the radiatively efficient QSO mode can have much effect in regulating galaxy growth and star formation shutdown, as postulated in the theoretical work cited above (Heckman 2010, Fabian 2012).

In the local Universe, such 'AGN or MBH feedback' has been observed in the so called 'radio mode' in very massive, central cluster galaxies driving jets into the intra-cluster medium. In these cases the central MBHs typically is in a fairly low or quiescent radiative state. Considerations of energetics suggest that radio mode feedback plausibly prevents cooling cluster gas to fall onto these massive galaxies that would otherwise lead to substantial further star formation and mass growth (McNamara & Nulsen 2007, Fabian 2012, Heckman & Best 2014). A second MBH feedback mode (termed 'QSO mode'), in which the MBH is active (i.e., the AGN is luminous) is detected as ionized winds from AGN (e.g. Cecil, Bland, & Tully 1990, Veilleux, Cecil & Bland-Hawthorn 2005, Westmoquette et al. 2012, Rupke & Veilleux 2013, Harrison et al. 2014) and from obscured QSOs (Zakamska & Greene 2014). The QSO mode feedback in form of powerful neutral and ionized gas outflows has also been found in late stage, gas rich mergers (Fischer et al. 2010, Feruglio et al. 2010, Sturm et al. 2011, Rupke & Veilleux 2013, Veilleux et al. 2013), which however are rare in the local Universe.

At high-z AGN QSO mode feedback has been seen in broad absorption line quasars (Arav et al. 2001, 2008, 2013, Korista et al. 2008), in type 2 AGN (Alexander et al. 2010, Nesvadba et al. 2011, Cano Díaz et al. 2012, Harrison et al. 2012), and in radio galaxies (Nesvadba et al.2008). However, luminous AGNs near the Eddington limit are again rare, constituting less than 1% of the star forming population in the same mass range (e.g. Boyle et al. 2000). QSOs have short lifetimes relative to the Hubble time ($t_{QSO} \sim 10^7 - 10^8$ yr $\ll t_H$, Martini 2004) and thus have low duty cycles compared to galactic star formation processes ($t_{SF} \sim 10^9$ yr, Hickox et al. 2014). It is thus not clear whether the radiatively efficient QSO mode can have much effect in regulating galaxy growth and star formation shutdown, as postulated in the theoretical work cited above (Heckman 2010, Fabian 2012).

From deep adaptive optics assisted integral field spectroscopy at the ESO VLT, Förster Schreiber et al. (2014) and Genzel et al. (2014) have recently reported the discovery of broad (($\sim 10^3$km/s), spatially resolved (a few kpc) ionized gas emission associated with the nuclear regions of very massive ($\log(M*/M_\odot) >$ 10.9) $z \sim 1-2$ star forming galaxies (SFGs). While active AGN do exhibit similar outflows, as stated above, the key breakthrough of this study is that it provides compelling evidence for wide-spread and powerful nuclear outflows in most ($\sim$ 70%) normal massive star forming galaxies at the peak of galaxy formation activity. The fraction of active, luminous AGN among this sample is 10-30%, suggesting that the nuclear outflow phenomenon has a significantly higher duty cycle than the AGN activity. If so, MBHs may indeed be capable to contribute to the quenching of star formation near the Schechter mass, as proposed by the theoretical work mentioned above.

8. Non-Merger Evolution Paths of MBHs

The most recent data on MBH demographics (Figure 4) suggest that the simple scenario of early MBH-galaxy formation through mergers and strong "co-evolution" might be too simplistic. Kormendy & Ho (2013, see also Kormendy, Bender & Cornell 2011) as well as McConnell & Ma (2013) find that MBHs in late type galaxies tend to fall below the best correlation of the pure spheroidal systems. The "pseudo"-bulges in these disk galaxies (including the Milky Way itself) typically rotate rapidly and may have partially formed by radial transport of disk stars to the nucleus mediated through slow, secular angular momentum transport, rather than by rapid merger events. In these systems the efficiency and growth processes of MBHs appears to be lower than in the very massive spheroids that formed a long time ago. In the local Universe, the Sloan Digital Sky Survey has shown that most AGN are not involved in active mergers or galaxy interactions (Li et al. 2008). Most lower luminosity AGN are in massive early type hosts that are not actively fed. Most of the lower-mass MBH growth at low redshift happens in lower mass galaxies (Kauffmann et al. 2003, Heckman et al. 2004).

Lower mass (between $10^{5.3}$ and 10^7 $M_\odot$) MBHs have been found in bulge-less disks and even dwarf galaxies (Filippenko & Ho 2003, Barth et al. 2001, Barth, Greene & Ho 2005, Greene & Ho 2004, 2007, Reines et al. 2011, 2013), in which there appears to be no or little correlation between the properties of the galaxy and its central MBH, in contrast to the bulged/spheroid systems (Greene 2012). These MBHs must have formed more through an entirely different path. MBH growth in these cases is more likely to be controlled by local processes, such gas infall from local molecular clouds (Sanders 1998, Genzel et al. 2010 and references therein) and stellar mass loss following a nuclear 'starburst' (Scoville & Norman 1988, Heckman et al. 2004, Davies et al. 2007, Wild et al. 2010).

At the peak of the galaxy formation epoch (redshifts $z \sim 1 - 2$) imaging studies show little evidence for the average AGNs to be in ongoing mergers (Cisternas et al. 2011, Schawinski et al. 2011, Kocevski et al. 2012). Instead most AGNs at this epoch are active star forming galaxies, including large disks, near the 'main-sequence' of star formation (Shao et al. 2010, Rosario et al. 2012, 2013). For active MBH, AGN luminosity and star formation rates are not or poorly correlated, excepting at the most extreme AGN luminosities (Netzer 2009, Rosario et al. 2012), yet the average MBH and galaxy growth rates are (Mullaney et al. 2012, del Vecchio et al. 2014). The empirical evidence for the AGN-merger model based on luminosity functions and spatial correlations (Hopkins et al. 2006) has been shown to not be a unique interpretation (Conroy & White 2013).

All these findings suggest that the concept of co-evolution between MBH growth and galaxy growth may most of the time be applicable only on average, or merely as a non- causal, statistical 'central limit' (Jahnke & Maccio 2011). One might call this 'weak' co- evolution. The instantaneous MBH growth rate at any given time exhibits large amplitude fluctuations (Hopkins et al. 2005, Novak et al. 2011, Rosario et al. 2012, Hickox et al. 2014). Relatively rare gas rich mergers may be able to stimulate phases of strong co- evolution at all redshifts. At other times, radial transport of gas (and stars) in galaxy disks may be an alternative channel of MBH growth, at least at the peak of galaxy-MBH formation, since galaxies 10 billion years were gas rich (Tacconi et al. 2013), resulting in efficient radial transport from the outer disk to the nucleus (a few hundred million years, Bournaud et al. 2011, Alexander & Hickox 2012). These inferences from the empirical data are in good agreement with the most recent hydrodynamical simulations (Sijacki et al. 2014).

9. MBH Spin

X-ray spectroscopy of the 6.4-6.7 keV Fe K-complex finds relativistic Doppler motions in several tens of AGNs, following the initial discovery in the iconic Seyfert galaxy MCG-6-30-15 (Tanaka et al. 1995, Figure 5). The Fe-K profiles can be modelled as a rotating disk on a scale of few to 20 R_S that reflects a power law, hard X-ray continuum emission component likely located above the disk (Tanaka et al. 1995, Nandra et al. 1997, 2007, Fabian et al. 2000, 2002, Fabian & Ross 2010, Reynolds 2013). While the X-ray spectroscopy by itself does not yield black hole masses, it provides strong support for the black hole interpretation. In addition, reverberation techniques of the time variable spectral properties are beginning to deliver interesting constraints on the spatial structure of the continuum and line components (Fabian et al. 2009, Uttley et al. 2014). From the modeling of the spectral profiles it is possible to derive unique constraints on the MBH spin, assuming that the basic modeling assumptions are applicable. The inferred spin for MCG-6-30-15 is near maximal (Figure 6). In a sample of 20 MBHs investigated in this way at least half have a spin parameter $a > 0.8$, providing tantalizing, exciting

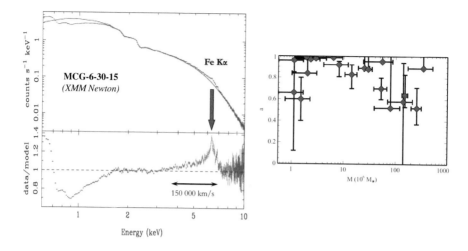

FIGURE 6. Left: Fe-K profile of the iconic Seyfert galaxy MCG-6-30-15 (Tanaka et al. 1995) obtained with XMM-Newton (Fabian et al. 2002, Fabian & Vaughan 2003). The blue extension of the relativistic emission extends as low as $\sim$ 3 keV. In the framework of a rotating accretion disk reflecting a hard X-ray power law component this means that there is emission at $\sim$ 2 × R_{grav}. If the MBH has a low angular momentum parameter, this would require significant reflection occurring within the last stable orbit. Alternatively, the MBH in MCG-6-3-15 is a near maximum Kerr hole with $a > 0.97$ (Brenneman & Reynolds 2006). Right: Inferred spin parameter a as a function of black hole mass, for 20 MBHs (Reynolds 2013).

evidence for a frequent occurrence of high-spin MBH (Figure 6, Reynolds 2013 and references therein). These measurements promise to yield important information on the growth processes of MBH.

10. Early Growth

The formation and evolution of MBHs faces two basic problems. One is angular momentum. To make it into the MBH event horizon from the outer disk of a galaxy, a particle has to lose all but 10^{-9} of its original angular momentum, a truly daunting task (c.f. Alexander & Hickox 2012). For this reason, major mergers have been considered natural candidate for being the sites of rapid MBH growth (Hopkins et al. 2006), since the mutual gravitational torques in a galaxy-galaxy interaction can reduce more than 90% of the angular momentum of a significant fraction of the total interstellar gas (e.g. Barnes & Hernquist 1996). However, this is by far not sufficient. It is likely that several stages of additional angular momentum loss much closer to the nucleus are involved in growing MBH, plausibly including star formation events at different 'way-points' including the nucleus itself (Scoville & Norman 1988, Davies et al. 2007, Hopkins & Quataert 2010, Meyer et al. 2010, Wild, Heckman & Charlot 2010, Alexander & Hickox 2012 and references therein). For this reason, black hole growth, accretion and radiation are probably highly time variable and strongly influenced by the properties of the gas and stellar environment in the sphere of influence around the black hole (Genzel et al. 2010, Hickox et al. 2014).

The second major obstacle is the time needed to grow to a final mass M from an initial seed of much lower mass M_0 (Volonteri 2010, 2012). This time is given by

$$\frac{t_M}{t_{\text{Salpeter}}} = \frac{\eta}{1-\eta} \times \frac{1}{L/L_{\text{Edd}}} \ln\left(\frac{M}{M_0}\right), \tag{1}$$

where $t_{\text{Salpeter}} = 4 \times 10^8$ yr, η is the radiative efficiency, L_{Edd} is the Eddington luminosity $(3.4 \times 10^4 M (\text{M}_\odot))$ where the accreting MBH's radiation pressure equals its gravity. To grow to $M = 10^9$ M$_\odot$ at $z \sim 6$ (~ 1 billion yr after the Big Bang) with $\eta \sim 0.1$ at the Eddington rate requires $4 \times 10^7 \ln(M/M_0)$ yr. If the initial seed formed in the re-ionization epoch at $z \sim 10$, the seed mass has to exceed $\sim 10^4$ M$_\odot$. While 10^9 M$_\odot$ MBHs as early as $z \sim 6$ are rare $(10^{-10} \text{Mpc}^{-3}$, Fan et al. 2006), and most very massive MBHs could have reached their final masses later, this example does show that standard Eddington accretion from a relative low mass seed, such as a super massive star $(M_0 \sim 10^2$ M$_\odot)$, cannot account for the oldest MBHs. Possibilities include fairly massive seeds $(\geq 10^4$ M$_\odot)$ formed from direct collapse of a dense gas cloud (Silk & Rees 1998), perhaps including a phase of super-Eddington accretion (see the more detailed account in Mitch Begelman's contribution in this volume, as well as the discussion in Begelman, Volonteri & Rees 2006, Volonteri 2010, 2012).

11. Zooming in on the Event Horizon

Looking forward to the next decade, there are several avenues to get still firmer constraints on the black hole paradigm, and determine the gravitational field still closer to the event horizon, in particular in the Galactic Center. Infrared spectroscopy of S2 during the next peri-approach in 2018 will have a good chance of detecting post-Newtonian parameters (Roemer effect, gravitational redshift, longitudinal Doppler effect, e.g. Zucker et al. 2006). Within the next decade it should be possible with current astrometric capabilities to detect S2's Schwarzschild precession angle, $\Delta\Phi_S = \frac{3\pi}{1-e^2}\left(\frac{R_S}{a}\right) \sim 12'$. The Schwarzschild precession and perhaps even the Lense-Thirring precession (due to the spin and quadrupole moment of the MBH) are obviously more easily detectable for stars with smaller semi-major axes and shorter orbital periods than S2. One such star, S102/S55 has been reported by Meyer et al. (2012), but the current confusion limited imagery on 10m-class telescopes prevents further progress.

This barrier will be broken in the next few years by the GRAVITY near-IR interferometric experiment (Eisenhauer et al. 2005a, 2011, Gillessen et al. 2006), and with the next generation 30m-class telescopes a decade later (Weinberg et al. 2005). GRAVITY will combine the four VLT telescopes interferometrically, with the goal of 10μarcsec precision, near-infrared imaging interferometry (angular resolution a few milli-arcseconds). GRAVITY will also be able to search for dynamical signatures of the variable infrared emission from SgrA* itself (Genzel et al. 2003, Eckart et al. 2006, Do et al. 2009, Dodds-Eden et al. 2009, c.f. Baganoff et al. 2001). These 'flares' originate from within a few milli-arcseconds of the radio position of SgrA* and probably occur when relativistic electrons in the innermost accretion zone of the black hole are substantially accelerated so that they are able to produce infrared synchrotron emission and X-ray synchrotron or inverse Compton radiation (Markoff et al. 2001). As such the infrared variable emission as well as the millimeter and submillimeter emission from SgrA* probe the inner accretion zone between a few to 100 RS. If orbital motion (of hot spots) could be detected the space time very close to the event horizon could potentially be probed (Paumard et al. 2005, Broderick & Loeb 2006, 2009, Hamaus et al. 2009).

VLBI at short millimeter or submillimeter wavelengths may be able to map out the strong light bending ('shadow') region inside the photon orbit of the MBH (Bardeen 1973, Falcke, Melia & Algol 2000). This 'Event Horizon Telescope' Project (http://www.eventhorizontelescope.org/) will soon benefit the observations of the Galactic Center (and the relatively nearby $\sim 6 \times 10^9$ M$_\odot$ MBH in M87) from the much enhanced sensitivity and additional u-v-coverage with the ALMA interferometer in Chile (Lu et al. 2014). It is hoped that the shadow signature can be extracted fairly easily even from data with a sparse coverage of the UV-plane (e.g. Doeleman 2010). As in the case of the GRAVITY observations of the infrared flares, it is not clear, however, how the potentially complex emission structure from the inner accretion zone, including a possible radio jet, may compromise the interpretation of EHT maps in terms of GR effects (Dexter & Fragile 2013, Moscibrodzka et al. 2014).

Given the current presence of $\sim$200 OB stars in the central parsec (Genzel et al. 2010) and extrapolating to earlier star formation episodes, there should be 100-1000 neutron stars, and thus potentially many pulsars within the parsec-scale sphere of influence of the Galactic Center MBH (Pfahl & Loeb 2004, Wharton et al. 2012). Until recently none have been found, despite many radio searches. The blame was placed on the large dispersion of the radio pulses by large columns of electrons in front of the Galactic Center sight line. In 2013 SWIFT and NUS-TAR discovered a magnetar, 1745-2900, within 3" of SgrA*, with a pulse period of 3.7 s, whose radio pulse characteristics have since been studied in detail (Kennea et al. 2013, Mori et al. 2013, Eatough et al. 2013, Spitler et al. 2014, Bower et al. 2014). While the magnetar itself cannot be used for timing studies, its detection renews the hope that radio pulsars can be detected with sufficient sensitivity but also increases the suspicion that there are physical germane to the Galactic Center region suppressing the formation of normal radio pulsars. If radio pulsars can be detected, however, precision timing, especially with the future capabilities of the Square Kilometer Array (SKA), has the potential to detect not only post-Newtonian parameters (including the Shapiro delay and the Schwarzschild precession term), but also the Lense-Thirring and quadrupole terms (Liu et al. 2012).

If one or several of these efforts is successful, it may be ultimately possible to test GR in the strong curvature limit, and test the no-hair theorem (Will 2008, Merritt et al. 2010, Psaltis & Johanssen 2011).

Acknowledgements

I would like to thank Stefan Gillessen, David Rosario, Amiel Sternberg, Scott Tremaine and Stijn Wuyts for helpful comments on this manuscript.

References

[1] Alexander, D. M., Swinbank, A. M., Smail, I., McDermid, R., & Nesvadba, N. P. H.: MNRAS **402**, 2211 (2010).

[2] Alexander, D.M., & Hickox, R.C.: New. Astr. Rev. **56**, 92 (2012).

[3] Almheiri, A., Marolf, D., Polchinski, J., & Sully, J.: JHEP **02**, 62 (2013).

[4] Arav, N., de Kool, M., Korista, K. T., et al.: ApJ **561**, 118 (200).

[5] Arav, N., Moe, M., Costantini, E., Korista, K. T., Benn, C., & Ellison, S.: ApJ **681**, 954 (2008).

[6] Arav, N., Borguet, B., Chamberlain, C., Edmonds, D., & Danforth, C.: MNRAS **436**, 3286 (2013).

[7] Argon, A.L., Greenhill, L.J., Reid, M.J., Moran, J.M., & Humphreys, E.M.L.: ApJ **659**, 1040 (2007).

[8] Baganoff, F. et al.: Nature **413**, 45 (2001).

[9] Baldry, I. K., Glazebrook, K., & Driver, S. P.: MNRAS **388**, 945 (2008).

[10] Bardeen J. M.: In Black holes (Les astres occlus), DeWitt B. S., DeWitt C., eds., New York: Gordon and Breach, p. 215, 1973.

[11] Barnes, J.E., & Hernquist, L. : ApJ **471**, 115 (1996).

[12] Barth, A.J., Sarzi, M., Rix, H.-W., Ho, L. C., Filippenko, A.V. & Sargent, W. L. W.: ApJ **555**, 685 (2001).

[13] Barth, A.J., Greene, J.E. & Ho, L.C.: ApJ **619**, L151 (2005).

[14] Begelman, M.C., Volonteri, M., & Rees, M.J.: MNRAS **370**, 289 (2006).

[15] Behroozi, P.S., Wechsler, R.H. & Conroy, C.: ApJ **770**, 57 (2013).

[16] Bender, R. et al.: ApJ **631**, 280 (2005).

[17] Bennert, V. N., Auger, M.W., Treu, T., Woo, J.-H., & Malkan, M.A.: ApJ **742**, 107 (2011).

[18] Bentz, M.C. et al.: ApJ **720**, L46 (2010).

[19] Blandford, R.D., & McKee, C.F.: ApJ **255**, 419 (1982).

[20] Blandford, R.D.: ASPC **160**, 265 (1999).

[21] Bournaud, F., Dekel, A., Teyssier, R., Cacciato, M., Daddi, E., Juneau, S., & Shankar, F.: ApJ **741**, L33 (2011).

[22] Bousso, R.: Rev. Mod. Phys. **74**, 825 (2002).

[23] Bower, G.C., et al.: Science **304**, 704 (2004).

[24] Bower, G.C., et al.: ApJ **780**, L2 (2014).

[25] Bower, R.G, Benson, A. J., Malbon, R., Helly, J. C., Frenk, C. S., Baugh, C. M., Cole, S. & Lacey, C. G.: MNRAS **370**, 645 (2006).

[26] Boyle, B. J., Shanks, T., Croom, S. M., Smith, R. J., Miller, L., Loaring, N., & Heymans, C.: MNRAS **317**, 1014 (2000).

[27] Braatz, J.A., Reid, M.J., Humphreys, E.M.L., Henkel, C., Condon, J.J., & Lo, K.Y.: ApJ **718**, 657 (2010).

[28] Brenneman, L.W., & Reynolds, C.S.: ApJ **652**, 1028 (2006).

[29] Broderick, A. E., & Loeb, A.: MNRAS **367**, 905 (2006).

[30] Broderick, A., Loeb, A., & Narayan, R.: ApJ **701**, 1357 (2009).

[31] Broderick, A. E., Fish, V. L., Doeleman, S. S., & Loeb, A.: ApJ **697**, 45 (2009).

[32] Cano-Díaz, M., Maiolino, R., Marconi, A., et al.: A&A **537**, L8 (2012).

[33] Cattaneo, A., Dekel, A., Devriendt, J., Guiderdoni, B., & Blaizot, J.: MNRAS **370**, 1651 (2006).

[34] Cecil, G., Bland, J., & Tully, B. R.: ApJ **355**, 70 (1990).

[35] Chatterjee, P., Hernquist, L., & Loeb, A.: ApJ **572**, 371 (2002).

[36] Chatzopoulos, S., Fritz, T., Gerhard, O., Gillessen, S., Wegg, C., Genzel, R., & Pfuhl, O.: arXiv1403.5266 (2014).

[37] Cisternas, M., et al.: ApJ **726**, 57 (2011).

[38] Conroy, C., & Wechsler, R.H.: ApJ **696**, 620 (2009).

[39] Conroy, C., & White, M.: ApJ **762**, 70 (2013).

[40] Croton, D.J., et al.: Mon. Not. Roy. Astr. Soc. **365**, 11 (2006).

[41] Davies, R.I., Müller Sánchez, F., Genzel, R., Tacconi, L. J., Hicks, E. K. S., Friedrich, S., & Sternberg, A.: ApJ **671**, 1388 (2007).

[42] Davis, T. A., Bureau, M., Cappellari, M., Sarzi, M., & Blitz, L.: Nature **494**, 328 (2013).

[43] Dekel, A., & Silk, J.: ApJ **303**, 39 (1986).

[44] Dekel, A., & Birnboim, Y.: MNRAS **368**, 2 (2006).

[45] DelVecchio, I., Lutz, D., Berta, S., et al.: In prep. (2014).

[46] Dexter, J., & Fragile, P.C.: MNRAS **432**, 2252 (2013).

[47] Di Matteo, T., Springel, V., & Hernquist, L.: Nature **433**, 604 (2005).

[48] Do, T., Ghez, A. M., Morris, M. R., Yelda, S., Meyer, L., Lu, J. R., Hornstein, S. D., & Matthews, K.: ApJ **691**, 1021(2009).

[49] Dodds-Eden, K., et al.: ApJ **698**, 676 (2009).

[50] Doeleman, S.S., et al.: Nature **455**, 78 (2008).

[51] Doeleman, S.S.: In Proceedings of the 10th European VLBI Network Symposium and EVN Users Meeting: VLBI and the new generation of radio arrays. September 20-24, 2010. Manchester, UK. Published online at http://pos.sissa.it/cgi-bin/reader/conf.cgi?confid=125, id.53, 2010.

[52] Eatough, R.P., et al.: Nature **501**, 391 (2013).

[53] Eckart, A., & Genzel, R.: Nature **383**, 415 (1996).

[54] Eckart, A., & Genzel, R.: MNRAS **284**, 576 (1997).

[55] Eckart, A., et al.: A&A **450**, 535 (2006).

[56] Efstathiou, G.: MNRAS **317**, 697 (2000).

[57] Einstein, A.: Ann. Phys. **49**, 50 (1916).

[58] Eisenhauer, F., et al.: 2003, ApJ **597**, L121 (2003).

[59] Eisenhauer, F., et al.: 2005, ApJ **628**, 246 (2005).

[60] Eisenhauer, F., Perrin, G., Rabien, S., Eckart, A., Lena, P., Genzel, R., Abuter, R., & Paumard, T.: Astr. Nachrichten **326**, 561 (2005a).

[61] Eisenhauer, F., et al.: ESO Msngr. **143**, 16 (2011).

[62] Fabian, A.C., Iwasawa, K., Reynolds, C.S., & Young, A.I.: PASP **112**, 1145 (2000).

[63] Fabian, A. C., Vaughan, S., Nandra, K., Iwasawa, K., Ballantyne, D. R., Lee, J. C., De Rosa, A., Turner, A., & Young, A. J.: MNRAS **335**, L1 (2002).

[64] Fabian, A.C., & Vaughan, S.: MNRAS **340**, L28 (2003).

[65] Fabian, A.C., et al.: Nature **459**, 540 (2009).

[66] Fabian, A.C., & Ross, R.R.: Sp. Sc. Rev. **157**, 167 (2010).

[67] Fabian, A. C.: 2012, ARA&A **50**, 455 (2013).

[68] Falcke, H., Melia, F., & Algol, E.: ApJ **528**, L13 (2000).

[69] Fan, X., et al.: 2006, AJ **132**, 117 (2006).

[70] Feruglio, C., Maiolino, R., Piconcelli, E., et al.: A&A **518**, L155 (2010).

[71] Ferrarese, L., & Merritt, D.: ApJ **539**, L9 (2000).

[72] Filippenko, A.V., & Sargent, W.L.W.: ApJ **342**, L11 (1989).

[73] Filippenko, A.V., & Ho, L.C.: ApJ **588**, L13 (2003).

[74] Fischer, J., Sturm, E., González-Alfonso, E., et al.: 2010, A&A **518**, L41 (2010).

[75] Förster Schreiber, N. M., Genzel, R., Newman, S. F., et al.: ApJ **787**, 38 (2014).

[76] Gebhardt, K., et al.: ApJ **539**, L13 (2000).

[77] Genzel, R., Hollenbach, D., & Townes, C. H.: Rep. Prog. Phys. **57**, 417 (1994).

[78] Genzel, R., Thatte, N., Krabbe, A., Kroker, H., & Tacconi-Garman, L.E.: ApJ **472**, 153 (1996).

[79] Genzel, R., Eckart, A., Ott, T., & Eisenhauer, F.: MNRAS **291**, 219 (1997).

[80] Genzel, R., Pichon, C., Eckart, A., Gerhard, O.E., & Ott, T.: MNRAS **317**, 348 (2000).

[81] Genzel, R., et al.: Nature **425**, 934 (2003).

[82] Genzel, R., Eisenhauer, F., & Gillessen, S.: 2010, Rev. Mod. Phys. **82**, 3121 (2010).

[83] Genzel, R., et al.: ApJ in press, (2014), (arXiv1406.0183).

[84] Ghez, A.M., Klein, B.L., Morris, M., & Becklin, E.E.: 1998, ApJ **509**, 678 (1998).

[85] Ghez, A.M., et al.: ApJ **586**, L127 (2003).

[86] Ghez, A., et al.: ApJ **620**, 744 (2005).

[87] Ghez, A., et al.: ApJ **689**, 1044 (2008).

[88] Giacconi, R., Gursky, H., Paolini, F., & Rossi, B.B.: Phys. Rev. Lett. **9**, 439 (1962).

[89] Giacconi, R.: Rev. Mod. Phys. **75**, 995 (2003).

[90] Gillessen, S.: SPIE **6268**, E11 (2006).

[91] Gillessen, S., et al.: ApJ **692**, 1075 (2009).

[92] Gillessen, S., et al.: ApJ **707**, L114 (2009a).

[93] Greene, J.E., & Ho, L.C.: ApJ **610**, 722 (2004).

[94] Greene, J.E., & Ho, L.C.: ApJ **667**, 131 (2007).

[95] Greene, J.E.: Nature Comm. **10.1038**, 2314 (2012).

[96] Gültekin, K., et al.: ApJ **698**, 198 (2009).

[97] Guo, Q., White, S., Li, C., & Boylan-Kolchin, M.: MNRAS **404**, 1111 (2010).

[98] Haehnelt, M.: In 'Coevolution of Black Holes and Galaxies', Carnegie Observatories Centennial Symposia. Cambridge University Press, Ed. L.C. Ho, p. 405, 2004.

[99] Häring, N., & Rix, H.-W.: ApJ **604**, L89 (2004).

[100] Haiman, Z., & Quataert, E.: In Supermassive Black Holes in the Distant Universe. Edited by Amy J. Barger, Astrophysics and Space Science Library Volume 308. ISBN 1-4020-2470-3 (HB), ISBN 1-4020-2471-1 (e-book). Published by Kluwer Academic Publishers, Dordrecht, The Netherlands, p. 147, 2004.

[101] Haller, J.W., Rieke, M.J., Rieke, G.H., Tamblyn, P., Close, L., & Melia, F.: ApJ **456**, 194 (1996).

[102] Hamaus, N., Paumard, T., Müller, T., Gillessen, S., Eisenhauer, F., Trippe, S., & Genzel, R.: ApJ **692**, 902 (2009).

[103] Harrison, C. M., Alexander, D. M., Swinbank, A. M., et al.: MNRAS **426**, 1073 (2012).

[104] Harrison, C. M., Alexander, D. M., Mullaney, J. R., & Swinbank, A. M.: MNRAS **441**, 3306 (2014).

[105] Heckman, T. M., Kauffmann, G., Brinchmann, J., Charlot, S., Tremonti, C., & White, S. D. M.: ApJ **613**, 109 (2004).

[106] Heckman, T.M.: In Co-Evolution of Central Black Holes and Galaxies, Proceedings of the International Astronomical Union, IAU Symposium, Volume 267, p. 3-14, 2010.

[107] Heckman, T.M., & Best, P.N.: ARAA (2014), in press (astro-ph 1403.4620).

[108] Herrnstein, J.T., Moran, J.M., Greenhill, L.J., & Trotter, A.S.: ApJ **629**, 719 (2005).

[109] Hickox, R. C., Mullaney, J. R., Alexander, D. M., Chen, C.-T. J., Civano, F.M., Goulding, A. D., & Hainline, K. N.: ApJ **782**, 9 (2014).

[110] Ho, L.C., Filippenko, A.V., & Sargent, W.L.W.: ApJS **112**, 315 (1997).

[111] Ho, L.C.: ARAA **46**, 475 (2008).

[112] Hopkins, P.F., et al.: ApJ **630**, 716 (2005).

[113] Hopkins, P.F.: ApJS **163**, 1 (2006).

[114] Hopkins, P.F., & Quataert, E.: MNRAS **407**, 1529 (2010).

[115] Jahnke, K., et al.: ApJ **706**, L215 (2009).

[116] Jahnke, K., & Maccio, A.V.: ApJ **743**, 92 (2011).

[117] Kaspi, S., Smith, P. S., Netzer, H., Maoz, D., Jannuzi, B. T., & Giveon, U.: ApJ **533**, 631(2000).

[118] Kauffmann, G., & Haehnelt, M.: MNRAS **311**, 576 (2000).

[119] Kauffmann, G., et al.: MNRAS **346**, 1055 (2003).

[120] Kennea, J.A., et al.: ApJ **770**, L24 (2013).

[121] Kerr, R.: Phys. Rev. Lett. **11**, 237 (1963).

[122] Kocevski, D.D., et al.: ApJ **744**, 148 (2012).

[123] Korista, K. T., Bautista, M. A., Arav, N., et al.: 2008, ApJ **688**, 108 (2008).

[124] Kormendy, J.: In 'Coevolution of Black Holes and Galaxies', Carnegie Observatories Centennial Symposia. Cambridge University Press, Ed. L.C. Ho, p. 1, 2004.

[125] Kormendy, J., Bender, R., & Cornell, M.E.: Nature **469**, 374 (2011).

[126] Kormendy, J., & Ho, L.: ARAA **51**, 511 (2013).

[127] Krabbe, A., et al.: ApJ **447**, L95 (1995).

[128] Kuo, C.Y., Braatz, J.A., Condon, J.J., et al.: ApJ **727**, 20 (2011).

[129] Lacy, J.H., Townes, C.H., Geballe, T.R., & Hollenbach, D.J.: ApJ **241**, 132 (1980).

[130] Lawson, P.R., Unwin, S.C., & Beichman, C.A.: JPL publication 04-014 (2004).

[131] Li, C., Kauffmann, G., Heckman, T. M., White, S. D.M., & Jing, Y. P.: MNRAS **385**, 1915 (2008).

[132] Liebling, S.L., & Palenzuela, C.: LRR **15**, 6 (2012).

[133] Liu, K., Wex, N., Kramer, M., Cordes, J. M., & Lazio, T. J. W.: ApJ **747**, 1 (2012).

[134] Lynden-Bell, D.: Nature **223**, 690 (1969).

[135] Lu, R.-S., et al.: ApJ **788**, L120 (2014).

[136] Macchetto, F., Marconi, A., Axon, D.J., Capetti, A., Sparks, W., & Crane, P.: ApJ **489**, 579 (1997).

[137] Madau, P., et al.: MNRAS **283**, 1388 (1996).

[138] Magorrian, J., et al.: AJ **115**, 2285 (1998).

[139] Maldacena, J.: Ad. Th. Math. Phys. **2**, 231 (1998).

[140] Maoz, E.: ApJ **447**, L91 (1995).

[141] Maoz, E.: ApJ **494**, L181 (1998).

[142] Marconi, A., et al.: ApJ **586**, 868 (2003).

[143] Marconi, A., Risaliti, G., Gilli, R., Hunt, L. K., Maiolino, R., & Salvati, M.: MNRAS **351**, 169 (2004).

[144] Marconi, A., Pastorini, G., Pacini, F., Axon, D. J., Capetti, A., Macchetto, D., Koekemoer, A. M., & Schreier, E. J.: A&A **448**, 921 (2006).

[145] Markoff, S., Falcke, H., Yuan, F., & Biermann, P.L.: Astr. & Ap. **379**, L13 (2001).

[146] Martini, P.: In Coevolution of Black Holes and Galaxies, from the Carnegie Observatories Centennial Symposia, Cambridge University Press, as part of the Carnegie Observatories Astrophysics Series, edited by L. C. Ho, p. 169, 2004.

[147] Mayer, L., Kazantzidis, S., Escala, A., & Callegari, S.: Nature **466**, 1082 (2010).

[148] McClintock, J., & Remillard, R.: In Compact Stellar X-ray sources , eds. W.Lewin and M.van der Klis, Cambirdge Univ. Press (astro-ph 0306123), 2004.

[149] McConnell, N., & Ma, C.-P.: ApJ **764**, 184 (2013).

[150] McGinn, M.T., Sellgren, K., Becklin, E.E., & Hall, D.N.B.: ApJ **338**, 824 (1989).

[151] McNamara, B.R., & Nulsen, P.E.J.: ARAA **45**, 117 (2007).

[152] Merloni, A., et al.: ApJ **708**, 137 (2010).

[153] Merritt, D., Alexander, T., Mikkola, S., & Will, C.M.: Phys. Rev. D **81**, 2002 (2010).

[154] Meyer, L., Ghez, A. M., Schödel, R., Yelda, S., Boehle, A., Lu, J. R., Do, T., Morris, M. R., Becklin, E. E., & Matthews, K.: Sci. **338**, 84 (2012).

[155] Mori, K., et al.: ApJ **770**, L23 (2013).

[156] Moscibrodzka, M., Falcke, H., Shiokawa, H., & Gammie, C. F.: arXiv:1408.4743 (2014).

[157] Moster, B. P., Naab, T., & White, S. D. M.: MNRAS **428**, 3121 (2013).

[158] Miyoshi, M., et al.: Nature **373**, 127 (1995).

[159] Moran, J.M.: ASPC **395**, 87 (2008).

[160] Mullaney, J.R., Daddi, E., Béthermin, M., Elbaz, D., Juneau, S., Pannella, M., Sargent, M. T., Alexander, D. M., Hickox, R. C.: ApJ **753**, L30 (2012).

[161] Munyaneza, F, Tsiklauri, D., & Viollier, R.D.: ApJ **509**, L105 (1998).

[162] Nandra, K., George, I. M., Mushotzky, R. F., Turner, T. J., & Yaqoob, T.: ApJ **477**, 602 (1997).

[163] Nandra, K., O'Neill, P. M., George, I. M.,& Reeves, J. N.: MNRAS **382**, 194 (2007).

[164] Nesvadba, N. P. H., Polletta, M., Lehnert, M. D., et al.: MNRAS **415**, 2359 (2011).

[165] Nesvadba, N. P. H., Lehnert, M. D., De Breuck, C., Gilbert, A. M., & van Breugel, W.: A&A **491**, 407 (2008).

[166] Netzer, H., & Peterson, B.M.: ASSL **218**, 85 (1997).

[167] Netzer, H., Mainieri, V., Rosati, P., & Trakhtenbrot, B.: A&A **453**, 525 (2006).

[168] Netzer, H.: MNRAS **399**, 1907 (2009).

[169] Netzer, H.: The Physics and Evolution of Active Galactic Nuclei, Cambridge University Press, 2013.

[170] Neumayer, N., Cappellari, M., Rix, H.-W., Hartung, M., Prieto, M. A., Meisenheimer, K., & Lenzen, R.: ApJ **643**, 226 (2006).

[171] Novak, G.S., Ostriker, J.P., & Ciotti, L.: ApJ **737**, 26 (2011).

[172] Osmer, P.S.: In Coevolution of Black Holes and Galaxies, from the Carnegie Observatories Centennial Symposia. Published by Cambridge University Press, as part of the Carnegie Observatories Astrophysics Series. Edited by L. C. Ho, p. 324, 2004.

[173] Özel, F., Psaltis, D., Narayan, R., & McClintock, J. E.: ApJ **725**, 1918 (2010).

[174] Paumard, T., Perrin, G., Eckart, A., Genzel, R., Lena, P., Schoedel, R., Eisenhauer, F., Mueller, T., & Gillessen, S.: Astr. Nachrichten **326**, 568 (2005).

[175] Peng, Y., et al.: ApJ **721**, 193 (2010).

[176] Peterson, B.M.: PASP **105**, 247 (1993).

[177] Peterson, B.M.: ASPC **290**, 43 (2003).

[178] Pfahl, E., & Loeb, A.: ApJ **615**, 253 (2004).

[179] Psaltis, D., & Johanssen, T.: JPhCS **283**, 2030 (2011).

[180] Rees, M.J., & Ostriker, J.P.: MNRAS **179**, 541 (1977).

[181] Rees, M.: Ann. Rev. Astr. Ap. **22**, 471 (1984).

[182] Reid, M.J., & Brunthaler, A.: ApJ **616**, 872 (2004).

[183] Reid, M.J., Braatz, J. A., Condon, J. J., Lo, K. Y., Kuo, C. Y., Impellizzeri, C. M. V., & Henkel, C.: ApJ **767**, 154 (2013).

[184] Reid, M.J., et al.: ApJ **783**, 130 (2014).

[185] Reines, A. E., Sivakoff, G. R., Johnson, K. E., & Brogan, C. L.: Nature **470**, 66 (2011).

[186] Reines, A.E., Greene, J.E., & Geha, M.: ApJ **775**, 116 (2013).

[187] Reynolds, C.S.: CQGra. **30**, 4004 (2013).

[188] Remillard, R.A., & McClintock, J.E.: ARAA **44**, 49 (2006).

[189] Rieke, G.H., & Rieke, M.J.: ApJ **330**, L33 (1988).

[190] Rosario, D., et al.: A&A **545**, 45 (2012).

[191] Rosario, D., et al.: ApJ **771**, 63 (2013).

[192] Rupke, D. S. N., & Veilleux, S.: ApJ **768**, 75 (2013).

[193] Sanders, D. B., et al.: ApJ **325**, 74 (1988).

[194] Sanders, R.H.: MNRAS **294**, 35 (1998).

[195] Schödel, R., et al.: Nature **419**, 694 (2002).

[196] Schödel, R., et al.: ApJ **596**, 1015 (2003).

[197] Schmidt, M.: Nature **197**, 1040 (1963).

[198] Schunck, F.E., & Mielke, E.W.: CQGra **20**, R301 (2003).

[199] Schawinski, K., Treister, E., Urry, C. M., Cardamone, C. N., Simmons, B., Yi, S. K.: ApJ **727**, L31 (2011).

[200] Schwarzschild, K.: Sitzungsber. Preuss. Akad.Wiss. **424**, (1916).

[201] Scoville, N.Z., & Norman, C.A.: ApJ **332**, 163 (1988).

[202] Sellgren, K., McGinn, M.T., Becklin, E.E., & Hall, D.N.: ApJ **359**, 112 (1990).

[203] Serabyn, E., & Lacy, J.H.: ApJ **293**, 445 (1985).

[204] Shakura, N.I., & Sunyaev, R.A.: A&A **24**, 337 (1973).

[205] Shankar, F., Weinberg, D.H., & Miralda-Escude, J.: ApJ **690**, 20 (2009).

[206] Shao, L., et al.: A&A **518**, L26 (2010).

[207] Shen, Z.Q., Lo, K.Y., Liang, M.C., Ho, P.T.P., & Zhao, J.H.: Nature **438**, 62 (2005).

[208] Sijacki, D., Vogelsberger, M., Genel, S., Springel, V., Torrey, P., Snyder, G., Nelson, D., & Hernquist, L.: arXiv1408.6842 (2014).

[209] Soltan, A.: MNRAS **200**, 115 (1982).

[210] Somerville, R., Hopkins, P. F., Cox, T. J., Robertson, B. E., & Hernquist, L.: MNRAS **391**, 481 (2008).

[211] Spitler, L.G., et al.: ApJ **780**, L3 (2014).

[212] Sturm, E., González-Alfonso, E., Veilleux, S., et al.: ApJ **733**, L16 (2011).

[213] Susskind, L.: JMP **36**, 6377 (1995).

[214] Tacconi, L.J., Neri, R., Genzel, R., et al.: ApJ **768**, 74 (2013).

[215] Tanaka, Y., et al.: Nature **375**, 659 (1995).

[216] Torres, D.F., Capoziello, S., & Lambiase, G.: Phys. Rev. D **62**, 4012 (2000).

[217] Townes, C. H., Lacy, J. H., Geballe, T. R., & Hollenbach, D. J.: Nature **301**, 661 (1982).

[218] Trakhtenbrot, B., & Netzer, H.: MNRAS **427**, 3081 (2012).

[219] Tremaine, S., et al.: ApJ **574**, 740 (2002).

[220] Uttley, P., et al.: A&A rev. **22**, 72 (2014).

[221] van der Marel, R.P., & van den Bosch, F.C.: AJ **116**, 2220 (1998).

[222] Vestergaard, M.: ApJ **601**, 676 (2004).

[223] Veilleux, S., Kim, D.-C., & Sanders, D.B.: ApJ **522**, 113 (1999).

[224] Veilleux, S., Cecil, G., & Bland-Hawthorn, J.: ARAA **43**, 769 (2005).

[225] Veilleux, S., et al.: ApJS **182**, 628 (2009).

[226] Veilleux, S., Meléndez, M., Sturm, E., et al.: ApJ **776**, 27 (2013).

[227] Viollier, R.D, Trautmann, D., & Tupper: PhLB **306**, 79 (1993).

[228] Volonteri, M.: A&AR **18**, 279 (2010).

[229] Volonteri, M.: Sci. **337**, 544 (2012).

[230] Weinberg, N. N., Milosavljevic, M., & Ghez, A. M.: ApJ **622**, 878 (2005).

[231] Westmoquette, M. S., Clements, D. L., Bendo, G. J., & Khan, S. A.: MNRAS **424**, 416 (2012).

[232] Wharton, R. S., Chatterjee, S., Cordes, J. M., Deneva, J. S., & Lazio, T. J. W.: ApJ **753**, 108 (2012).

[233] Wheeler, J.A.: Amer. Scient. **56**, 1 (1968).

[234] Wild, V., Heckman, T., & Charlot, S.: MNRAS **405**, 933 (2010).

[235] Will, C.M.: ApJ **674**, L25 (2008).

[236] Wollman, E. R., Geballe, T. R., Lacy, J. H., Townes, C. H., Rank, D. M.: ApJ **218**, L103 (1977).

[237] Yu, Q., & Tremaine, S.: MNRAS **335**, 965 (2002).

[238] Zakamska, N.L., & Greene, J.E.: MNRAS **442**, 784 (2013).

[239] Zucker, S., Alexander, T., Gillessen, S., Eisenhauer, F., & Genzel, R.: ApJ **639**, L21 (2006).

Reinhard Genzel

Max-Planck Institut für Extraterrestrische Physik (MPE)

Garching, Germany

and

Departments of Physics & Astronomy

University of California

Berkeley, USA

e-mail: `genzel@mpe.mpg.de`

The Universe, 121–147

New Worlds Ahead: The Discovery of Exoplanets

Arnaud Cassan

Abstract. Exoplanets are planets orbiting stars other than the Sun. In 1995, the discovery of the first exoplanet orbiting a solar-type star paved the way to an exoplanet detection rush, which revealed an astonishing diversity of possible worlds. These detections led us to completely renew planet formation and evolution theories. Several detection techniques have revealed a wealth of surprising properties characterizing exoplanets that are not found in our own planetary system. After two decades of exoplanet search, these new worlds are found to be ubiquitous throughout the Milky Way. A positive sign that life has developed elsewhere than on Earth?

Keywords. Astronomy, exo-planetary systems, exoplanets, extra-solar planets, gravitational microlensing, transits, radial velocity, direct imaging.

1. The Solar system paradigm: the end of certainties

Looking at the Solar system, striking facts appear clearly: all seven planets orbit in the same plane (the ecliptic), all have almost circular orbits, the Sun rotation is perpendicular to this plane, and the direction of the Sun rotation is the same as the planets revolution around the Sun.

These observations gave birth to the Solar nebula theory, which was proposed by Kant and Laplace more that two hundred years ago, but, although correct, it has been for decades the subject of many debates. In this theory, the Solar system was formed by the collapse of an approximately spheric giant interstellar cloud of gas and dust, which eventually flattened in the plane perpendicular to its initial rotation axis. The denser material in the center collapsed further under self-gravity, increased in density and formed the Sun. Outside, the material had collapsed into a disk-shape nebula, a gaseous flattened disk in differential rotation, where planets were supposed to form. An argument for that was that without planets orbiting the central star, the star's angular momentum would be so high that it would be disrupted by its own rotation speed, so in a sense planets were required.

Planets were therefore considered by a number of authors (notably Giordano Bruno) as by-products of a the global process of star formation, and the plurality of worlds was inferred as a natural consequence of the plurality of stars. First evidences that planet formation takes place in a proto-planetary disk composed of gas and dust were supported by observations of flux excess in the infrared and ultraviolet wavelengths, attributed to nebulae surrounding stars.

Another important fact about the Solar system lies in the arrangement of the planets around the Sun: the small, rocky planets Mercury, Venus, Earth and Mars are located at small orbital distances (less than 1.5 the Sun-Earth distance, or AU), gas giant planets like Jupiter (318 Earth masses, or $M_\oplus$) and Saturn ($95\,M_\oplus$) are located further away (5 and 10 AU), and finally the icy giants Uranus and Neptune, located even further (20 and 30 AU), are much less massive ($14\,M_\oplus$ and $17\,M_\oplus$). Pluto does not fit well in this picture, but in fact it is not considered anymore as a planet – it is more a heavy version of an asteroid, today classified as a dwarf planet, with a very different history.

A relatively simple planet formation theory can explain most of these characteristics, and was widely accepted amongst the scientific community until the first exoplanet was detected in 1995. To grow large gaseous planets, it is necessary that there is enough material available around. This requires that the orbit is large, because the amount of material inside the feeding ring of the planet is higher, and because the temperature is low enough to allow the condensation of ices, which increases the solid fraction of the material and ease the nucleation process. Jupiter and Saturn are in fact located beyond the snow line, which delimits the orbit at which most ices condense (water, methane,...). Orbits which are too far away suffer from the fact planets take more time to accomplish a given number of orbits, and thus accrete less gas before it falls in the star or is dispersed away by the stellar wind. Saturn is less massive than Jupiter, Uranus and Neptune are even lighter and contain a high fraction of ices. Small planets are found in relatively close orbits, because they have little material to accrete. All pieces of the puzzle made sense: Solar systems planets were formed *in situ*, at the orbital position they are seen today.

The paradigm that emerged from the observation of the Solar system was believed to also apply to other exo-planetary systems. The ideas behind it made the basis of the design of the first campaigns of exoplanet searches: find a Jupiter-like planet located a few AU from another star than the Sun. But the first exoplanet detected around a star similar to the Sun, 51 Peg b, was definitely not of that kind: although not different in nature from Jupiter (it was a gas giant), its orbit of only 4 days period against 12 years for Jupiter caused a great surprise: it was a *hot* Jupiter! The formation of this exoplanet *in situ* was clearly ruled out: too hot, not enough material.

Hence the only possibility was that the planet had formed at a larger orbital distance, and eventually migrated inward to get close to the star via an adequate mechanism. It finally stopped at its current location and narrowly avoided to be swallowed by the star. The discovery of this unexpected hot Jupiter immediately

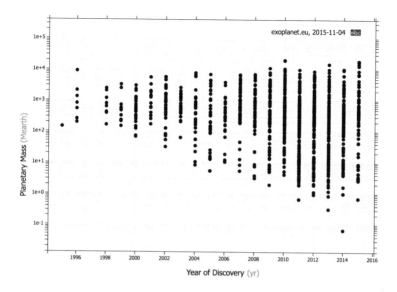

FIGURE 1. Exoplanet detections as a function of time since the discovery of 51 Peg b in 1995. The vertical axis displays the planetary mass in Earth masses, and shows the great improvements towards low masses with years, today down to that of the Earth (figure created with exoplanet.eu).

generated a feverish research activity: those who did not believe in the planetary interpretation worked out new stellar pulsation theories; others investigated theoretical and numerical scenarios for planetary migration in the proto-planetary disk, and actually re-discovered calculations made at the turn of the 1970s.

Today, not only the classical Solar system formation scenario described above cannot accommodate the discovery of exoplanets, but the history of the Solar system itself has undergone significant changes. One of such popular theory is the Grand Tack model (or Nice model) which proposes that Jupiter migrated inward to 1.5 AU from an initial 5 AU formation orbit, and then migrated back outward due to disk torques before and after Saturn's formation. This scenario can account for another important aspect: the delivery of water on Earth (and other terrestrial planets) in the form of water-rich planetesimals (today still present in the asteroid belt as water-rich asteroids) scattered inward during the gas giants' outward migration.

The first exoplanet detections have triggered an unprecedented rush to detect exoplanets (Fig. 1), which provide essential (indispensable) information to understand their great diversity or their physical properties. We still make amazing discoveries 20 years after the first detection.

2. Searching for exoplanets

The detection of extrasolar planets has always been a great observational challenge, because the angular separation between the planet and the host star is extremely small and because the brightness contrast is extremely high. At the beginning of the 1990s, the only example of a planetary system was our Solar system. The first exoplanets have been discovered by indirect methods. In 1992, the timing of the millisecond pulsar PSR1257+12 led to the discovery of planetary-mass objects around a neutron star. A few years later, the first exoplanet orbiting a Sun-like star, 51 Peg b, was discovered by high-accuracy radial velocity measurements of the star's periodic motion. These two landmark discoveries have initiated a novel, very active field in astrophysics: the search and characterization of extrasolar planets. At the end of 2013, the exoplanet catalogue passed the symbolic 1000th entry. The number of detections has reached almost 2000 detections today.

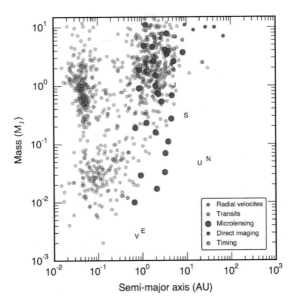

FIGURE 2. Exoplanet detections as of September 2015 with the main techniques described in the text, in a mass vs. semi-major axis diagram (Fig. courtesy C. Ranc).

The search for extra-solar planets have unveiled a striking fact: the great diversity of their physical properties. To introduce the main detection methods that are described in the coming paragraphs, the discoveries as of September 2015 are shown in the mass vs. semi-major axis diagram shown in Fig. 2.

2.1. Pulsar timing

In 1967, J. Bell and A. Hewish discovered in the sky the first member of a new kind of radio pulsating point-like sources that were called "pulsating stars", or pulsars. It soon after became clear, however, that these objects had little to do with real stars, but that they were rapidly rotating and highly magnetized neutrons stars, whose existence was then subject to speculation. A neutron star is the remnant of a very massive star after it exploded as a supernova. While the original enveloppe of the star is blown away, its core collapses very fast and the pressure becomes so high that protons and electrons cannot resist it and merge to form neutrons. The equivalent of the mass of the Sun is contained inside a sphere of only 20 km in diameter. A pulsar is a magnetized neutron star which emits powerful radio waves in two cone-shaped beams which are inclined with respect to the spin axis. Every time a cone points toward the Earth, a pulse is received. Their typical periods range from milliseconds to seconds.

It was soon realized that the pulsation period of the pulsars was intrinsically extremely stable (millisecond pulsars did actually serve as time references). This property was first used in 1974 by J. Taylor and R. Hulse who showed indirectly that PSR1913+16 (a binary neutron star including a pulsar) emitted gravitational waves as predicted by the theory of General Relativity: the gravitational loss of energy shrinks the orbits of the companions, which in turn shortens the period of the pulses. Pulsars surveys were subsequently carried out with increasingly large radio telescopes.

In early 1990, the routine operations at the Arecibo radio telescope were shut down for repairs of the damages caused by material fatigue that had developed over time. The astronomer A. Wolszczan took this opportunity to propose a large survey to discover new pulsars and probe the distribution of old neutron stars over the sky. The limited access to the telescope during the reparation phase made it practically unavailable to outside observers, and a large amount of time was granted to conduct his project. After a few months of monitoring, two new pulsars were found: one was part of a binary neutron star, while the second one, PSR B1257+12, was a millisecond pulsar with a spin period of 6.2 ms. The timing model of the latter did not fit well that of an isolated rotating neutron star, though. After several unsuccessful months spent in trying to refine the model, mid-1991, the pulsar was monitored during three weeks on a daily basis in order to track down the details of the discrepancy between the timing prediction and the actual observations. The pulse arrival times were found to trace a smooth curve (upper panel of Fig. 3), which was finally interpreted as a sign of a periodic phenomenon affecting the pulsar.

If the pulsar was perturbed by an orbiting companion such as a white dwarf (also a stellar remnant, but for a least massive progenitor star) as it was *a priori* the most likely interpretation, its reflex motion should translate into Doppler shifts of the apparent pulsar period. For a single Keplerian orbit, the varying delay Δt_R

between pulses is given by

$$\Delta t_R = x(\cos E - e)\sin\omega + x\sin E\sqrt{1 - e^2}\cos\omega\,, \qquad (1)$$

where $x = (a\sin i)/c$, c is the speed of light, a is the semi-major axis, i is the orbital inclination, e is the eccentricity, E is the eccentric anomaly related to the mean anomaly $M = (2\pi/P)(t - t_p)$ through $M = E - e\sin E$, P is the orbital period, t_p is the time of the periastron passage and ω is the periastron longitude. The amplitude of the variation, however, implied a companion of terrestrial mass several order of magnitudes smaller than the mass of a white dwarf (pulsar timing is so sensitive that even asteroid-mass bodies are detectable, equivalent to $1\,\mathrm{cm.s}^{-1}$ in Doppler precision).

The detailed analysis finally revealed the presence of two terrestrial planets (about three and four times the mass of the Earth) around pulsar PSR B1257+12 at the time of the publication in 1992 [30]. This configuration provided a quasi-perfect fit the varying delay between the pulses, as can be seen in the lower panel of Fig. 3, which exhibits almost no residuals for the initial 18 months of data. Later in 1994, a third, Moon-mass planet was found to orbit the pulsar too, at a closer orbit.

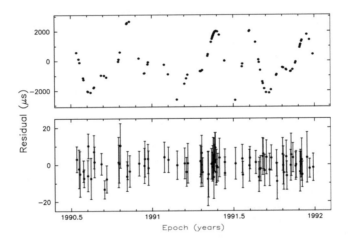

FIGURE 3. The upper panel shows the residuals of the pulse arrival times for the best pulsar timing model without companion (large residuals), while the lower panel is the same as above but including two companion terrestrial planets (figure from [30]).

Hence, and very unexpectedly, the first planetary-mass objects have been found around a stellar remanent (sometimes called dead star), and not around a normal star as it was commonly expected. Nevertheless, the possible existence of planets around pulsars had been investigated shortly after the first pulsar discoveries [18], and two pulsar-planets were announced and later retracted before the discovery of PSR B1257+12. The story says that the announcement of the detection of planets around pulsar PSR1829-10 [3] and around PSR B1257+12 were programmed at the same conference, but the authors finally retracted their claim in public just before the announcement of the PSR B1257+12 planets. A second pulsar-planet system was finally discovered two years later [2].

The discovery of planets around pulsars provide a strong support that exoplanets exist around stars at all stages of their evolution, even in their final ones. Whether these planets have been formed from the fallback accretion of matter left in a post-supernova debris disk, as suggested by the quasi-planar architecture of the three planets around PSR B1257+12, or whether (although it is less likely) they are objects that have survived the explosion is not clear yet. The detection so far of only two planetary systems around millisecond pulsars means that building up planets around pulsars is not a common process, but it also supports the idea that the formation of terrestrial planets is an efficient process even in unfavorable environments.

2.2. Doppler spectroscopy

The first exoplanet orbiting a normal star (i.e., a star burning mainly hydrogen as its source of heat, like the Sun) was discovered in 1995 by Doppler spectroscopy of its host star. It has remained the most productive detection technique for about 20 years, before space missions dedicated to transiting planet search took the lead in terms of number of detections. This technique requires a very accurate spectrograph which measures the periodic Doppler-Fizeau shift of the star's spectra as it moves around the star-planet barycenter. This wavelength shift $\Delta\lambda$ is then translated into a measurement of the radial velocity v_r of the star towards the observer through $\Delta\lambda/\lambda = -v_r/c$. A Doppler precision of $1\,\mathrm{m.s^{-1}}$ typically corresponds to a stellar lines shift of 1/1000th of a CCD pixel.

In practice, the measured semi-amplitude K_* of the radial velocity of the host star $v_r(t)$ can be expressed as

$$K_*[\mathrm{cm\,s^{-1}}] = \frac{8.95}{\sqrt{1-e^2}} \frac{m\sin i}{\mathrm{M_\oplus}} \left(\frac{M_*+m}{\mathrm{M_\odot}}\right)^{-2/3} \left(\frac{P}{\mathrm{yr}}\right)^{-1/3}, \qquad (2)$$

where m is the planet mass, M_* is the star mass, $\mathrm{M_\odot}$ and $\mathrm{M_\oplus}$ are the Sun and Earth masses respectively, P is the orbital period expressed in years, e is the orbit eccentricity and i is the orbit inclination. The detailed modeling of the radial velocity curve $v_r(t)$ yields the measurement of P, as well as the eccentricity (distortion of $v_r(t)$ relative to a sinusoidal curve), the longitude and time of the passage at the periastron, and of K_*. But as seen in the expression of K_*, the true mass of the planet m and the inclination i of the orbit remain degenerated. Hence only the planet minimum mass $m\sin i$ is measured. Statistically, the probability that the inclination lies within $i_1 < i < i_2$ is given by $P = |\cos i_2 - \cos i_1|$. It means that for example, there is 87% probability that the inclination of a given planet lies between thirty and ninety degrees (pure radial motion), or equivalently, a 87%

probability that the true mass lies between the measured $m \sin i$ and twice this value.

The reflex motion that Jupiter exerts on the Sun is about $K_* \sim 12.5 \, \mathrm{m.s^{-1}}$, while it drops to $\sim 0.09 \, \mathrm{m.s^{-1}}$ when considering the pull of the Earth. These values have to be compared to the typical radial velocity precision achieved by the spectrographs: while in 1995, it was about $10 \, \mathrm{m.s^{-1}}$, in 1998 it improved to $3 \, \mathrm{m.s^{-1}}$ and reached $1 \, \mathrm{m.s^{-1}}$ in 2005 when the HARPS instrument mounted on the 3.6m telescope in La Silla (ESO Chile) was commissioned. These values may explain that there is so little literature speculating about the possibility of detecting exoplanets by measuring the radial velocity of stars before the first detection. The most likely reason is that the first generation of spectroscopes were far from being accurate enough and did not allow much hope. Indeed, measuring a Doppler shift is a very challenging task that requires high signal-to-noise ratio, high resolution, and large spectral coverage. The use of photographic plates and the approximate guiding at the spectrograph slit in the early 1970s limited the sensitivity to accuracies of about $1 \, \mathrm{km.s^{-1}}$. The advent of echelle spectrometers (using high diffraction orders) have revolutionized Doppler spectroscopy and allowed to reach the required precision to detect brown dwarfs and even exoplanets.

The formula giving K_* shows that more massive planets are easier to detect (K_* increases with m), as well as shorter period (i.e., close-in) planets. Planets are also easier to find around low-mass stars than heavier stars. Furthermore, a planet must at least complete one full orbit in order to have its parameters constrained (although more orbits are usually needed to obtain good constraints). Hence when more data are collected with time, planets on larger orbits become detectable, in particular additional planets in already discovered systems.

Back in the early 1990s, and taking the Solar system as a reference, the short-orbit planets (Venus, Earth or Mars) were not massive enough to be detected, and the detection of Jupiter would require to wait for about 12 years. Early radial velocity searches were actually mainly focused on the characterization of the substellar and brown dwarf mass function by searching for companions of main sequence stars below one solar mass [12, 21], or were dedicated to establish improved radial velocity standards. As the accuracies improved toward $10 - 20 \, \mathrm{m.s^{-1}}$, the efforts increased and new observing programs started, leading to the monitoring of many more stars. At the end of the 1980s, many claims of exoplanet detections were retracted, which progressively introduce skepticism in the field. The case of γ Cep [6] is an instructive example: in 1988, variations in the residual velocities were clearly identified, but were attributed to stellar activity. In 2003, a reanalysis of the data obtained between 1981-2002 finally confirmed a planetary signal fifteen years later. Similarly in 1989, a 84-days periodic Doppler signal was detected around the star HD 114762, implying a companion of minimum mass of 11 Jupiter masses [20]. But because of the ambiguity on the true mass since the orbital inclination was unknown, the data were misinterpreted as a probable brown dwarf signal, and only confirmed as a planet some years later.

In the climate of suspicion that dominates the mid-1990s, when M. Mayor et D. Queloz announced the detection of a possible planet around the solar-like star 51 Peg at the Observatoire de Haute Provence, stellar oscillations or non-radial pulsations were immediately invoked as possible sources of confusion in several publications. But the most striking fact about the discovery resided in the fact that the best model implied a minimum mass of about half of that of Jupiter, with an extraordinary orbital period of only 4.2 days (Fig. 4). For comparison, in the Solar system Jupiter has an orbital period of 12 years. While this very unexpected claim could have made an easy argument to refute the planetary interpretation, a number of events turned the situation around. First, the discovery was promptly confirmed by the Lick Observatory group, and second, this same group was also able to report two additional similar planets (large mass and very short orbits) around the Sun-like stars 70 Vir [22] and 47 UMa [5]. It then became quite clear that the main reason exoplanets were missed in the early years of monitoring is that the surveys were dedicated to detect planet with orbital periods larger than 10 years, where Jupiter analogs were thought to form. The discovery of these planets, called *hot Jupiters*, marked the beginning of a Doppler planet detection rush, and the birth of the field of extrasolar planet.

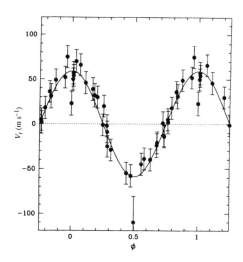

FIGURE 4. Radial velocity curve folded in phase of the 51 Peg b hot Jupiter. The model curve (solid line) accounts for a planet of half a Jupiter mass on a 4.2 days orbit (figure from [23]).

Nevertheless, the detection of significant eccentricity in many new exoplanet candidates did cast doubt again on the planetary interpretations, because many astronomers argued that planet should reside on circular orbits, in contrast with binary stars. Born in a gaseous disk, it was argued that planet eccentricities should be damped by the gas during the formation process. It is a well-known fact today that planets can have large eccentricities, but the doubt was not left until the first transiting planet and multi-planetary systems were detected just before the turn of the 21st century.

Doppler spectroscopy plays an important role to confirm transiting exoplanets. New designs using near-infrared spectroscopy allows to monitor low-mass stars (red dwarfs), which account for up to 80% of the number of stars in the Milky Way. The smaller mass of the star makes it easier to detect low mass planets and at larger orbits. Observations at larger wavelengths are furthermore less sensitive to stellar activity that can mimic planetary signals.

2.3. Transit

When an exoplanet passes in front of its star, and given a suitable alignment between the planet, the star and the observer, the light from the host star is decreased by the transit of a planet across its disk, with the effect repeating at the orbital period (Fig. 5). The phenomenon is similar to the transit of Venus in front of the Sun as has been recently observed from Earth. The first exoplanet transit was detected in HD 209458 [17, 9], which was already known to harbor a planet thanks to Doppler spectroscopy. The duration of the transit was about 2.5 hours and had a depth of about 1.5%.

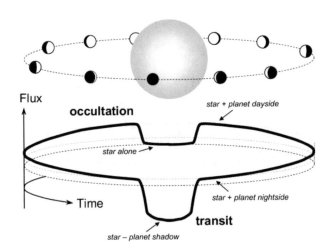

FIGURE 5. Transit curve with showing the primary eclipse (transit), the secondary eclipse (occultation) and the different phases (figure from [29]).

The probability that a given exoplanet transits its host star is primarily a function of the inclination of the planetary orbit and of the stellar radius,

$$P_{tr} = 0.0045 \left(\frac{\text{AU}}{a} \right) \left(\frac{R_\star + R_p}{R_\odot} \right) \left[\frac{1 + e \cos(\pi/2 - \omega)}{1 - e^2} \right], \tag{3}$$

where ω is the angle at which orbital periastron occurs ($\omega = 90°$ indicates transit), and e is the orbital eccentricity. A typical hot Jupiter around a solar-type star of radius $\sim R_{\mathrm{Jup}}$ and period $P \sim 3\,\mathrm{d}$ has a transit probability $\mathcal{P} \sim 10\%$, a transit duration of $\tau \sim 3\,\mathrm{hr}$ and a photometric transit depth of $d \sim 1\%$. A super-Earth (few Earth masses) has typically $\mathcal{P} \sim 2.5\%$, $d \sim 0.1\%$, and $\tau \sim 6\,\mathrm{hr}$, while an Earth-like planet at 1 AU around a solar-type stars has $\mathcal{P} \sim 0.5\%$, $d \sim 0.01\%$, and $\tau \sim 15\,\mathrm{hr}$, which makes the detection very challenging. For nearly a decade, the community as a whole struggled to implement productive surveys. Expectations in terms of planet yield were largely overestimated. But after good strategies and new instruments were designed in the few years following the first discoveries, progress in transiting exoplanet detections went very fast. This is thanks to the increasing number of ground-based projects, such as the SuperWASP and HATNet surveys. Dozens of transiting planets with high-quality light curves have been gathered with accurate masses determined with precision Doppler velocity measurements. Thousands of additional transiting planetary candidates have been observed from space, with space missions CoRoT (CNES) and Kepler (NASA). Transit timing variations have progressed from a theoretical exercise to a practiced technique.

A serious challenge for wide-field surveys lies in the many ways transit can be affected by astrophysical false positives. Radial velocity measurements is then required (although other techniques are possible) to confirm the planetary nature of the signal. In that case, the exact mass and radius of the planet are measured, which yield the mean density. False positives may have different origin. Low mass stars and brown dwarfs overlap in size with giant planets, and have almost identical transit signatures to those of giant planets. Grazing eclipsing binaries (only a small fraction of the star's disk is eclipsed by the companion) can also provide an important source of confusion for low signal-to-noise light curves. When an eclipsing binary, either physically related or unrelated, shares the line of sight with the target star (flux received on the same pixel area), the total flux is increased and the apparent relative transit depth is consequently decreased, so that the transit depth appear smaller. False alarm probabilities are inferred to be dramatically lower for cases where multiple planets transit the same star. In this case, exoplanets can be confirmed on the grounds of transit signals only. In practice for the Kepler survey, the majority of the candidate planets cannot be confirmed by Doppler measurements, but about 800 planets were confirmed thanks to their multiple planet transits.

The flagship space missions Kepler and CoRoT have both exhibited excellent productivity, and a third mission, MOST, has provided photometric transit discoveries of several previously known planets. The Doppler technique, which was by far the most productive discovery method through 2006, is rapidly transitioning from a general survey mode to an intensive focus on low-mass planets and to the characterization transiting planets. The Kepler mission did provide hundreds of planet detections with mass determinations, as well as hundreds of multiple transiting planets orbiting a single host star, many of which are coplanar and rather crowded systems. The CoRoT satellite ceased active data gathering in late

2012, having substantially exceeded its three-year design life. In 2013, the Kepler satellite experienced a failure of a second reaction wheel, which brought its high-precision photometric monitoring program to an end, after four years of delivering high precision transit data.

New transit space missions have already been programmed for the next 10 years. NASA has selected the TESS mission that is scheduled for launch in 2017. It will monitor the all sky to locate transiting planets with periods of weeks to months, and sizes down to $\sim 1\,R_\oplus$ among a sample of 5×10^5 stars including ~ 1000 red dwarfs. Another mission, the ESA CHEOPS satellite, is also scheduled for launch in 2017. It will selectively and intensively search for transits by already discovered high-precision Doopler planet candidates with radii in the range $1 - 4\,R_\oplus$. It will also perform follow-up observations of interesting TESS candidates. Finally, the ESA PLATO mission will take over in 2025, with the objective to find and study a large number of extrasolar planetary systems, with emphasis on the properties of terrestrial planets in the habitable zone around solar-like stars. The satellite has also been designed to investigate seismic activity in stars, enabling the precise characterization of the planet host star, including its age.

2.4. Gravitational microlensing

In 1936, Einstein derived the equations of the bending of light rays originating from a background star when passing in the vicinity of a foreground star, what is called today gravitational microlensing. At the time the article came out, however, observational facilities were not developed enough yet to seriously consider detecting a microlensing effect. Einstein himself concluded: "there is no great chance of observing this phenomenon". But 50 years after Einstein's publication, the astrophysicist Bohdan Paczyński [24] revisited the basic ideas of microlensing observations, in a seminal article published in 1986. The original idea of the paper was to propose a new method to detect hypothetic dark, compact, massive halo objects (MACHOs) as a possible form of dark matter in the Milky Way. A number of observational searches with line of sights towards the Large and Small Magellanic Clouds and the Galactic center in particular were subsequently initiated beginning of the 1990s: MACHO (Massive Compact Halo Object), EROS (Experience pour la Recherche d'Objets Sombres) and OGLE (Optical Gravitational Lensing Experiment). In 1993, the first stellar microlensing events were detected independently by the MACHO and EROS collaborations. These detections mark the birth of microlensing as an observational technique. About 3000 events are now detected every year, which provide unique astrophysical informations in several fields of research in astronomy and astrophysics.

Compared to other planet detection methods, microlensing detections bring unique information on planetary populations that justified the strong and steady efforts to make the technique work. While most planets detected with other methods are detected close to their stars, prime targets of microlensing are planets located beyond the snow line of their stars, where ices can start to form. A full understanding of the demographics of extrasolar planet in the Galaxy thus relies on the combination of the different observational techniques. During decade 2003-13, the most important microlensing results include the discovery of the first ever cool super-Earth planet (2005), the discovery of Jupiter-mass free-floating planets (2012), and first constraints on the planetary mass function for a wide range of masses and orbital distances (2012). At the end of a decade filled with discoveries, microlensing observations find that, on average, every Milky Way star has a planet, and that planets around stars are the rule rather than the exception.

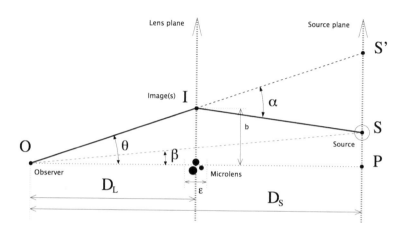

FIGURE 6. Geometry of a microlensing event. The source is located in the background, and is lensed by a foreground lensing star or planetary system (figure from [7]).

Gravitational microlensing describes the bending of light from a background source star due to the gravitational field of a compact object crossing the observer-source line-of-sight, and acting as a lens. The geometry of the problem is shown in Fig. 6. Light rays passing in the vicinity of the microlens will be bent by gravity by an angle $\hat{\alpha} \ll 2\pi$ given by

$$\hat{\alpha} = \frac{4GM}{c^2} \frac{1}{|b|} \, , \qquad (4)$$

where $|b|$ is the closest approach distance of the light ray from the lens, M is the total mass of the lens, c the speed of light and G the gravitational constant. In general, multiple images of the source are produced, but in the ideal case of a single point-mass lens perfectly aligned with a point-source, the image of the source is a circle of angular radius θ_E, or angular Einstein ring radius,

$$\theta_E = (\kappa M \pi_{\text{rel}})^{1/2} \, , \qquad (5)$$

where $\pi_{\text{rel}} \equiv \text{AU}/D_L - \text{AU}/D_S$ is the relative lens-source parallax, D_L and D_S are the observer-lens and observer-source distances, $\kappa \simeq 4G/c^2\text{AU} \simeq 8.144\,\text{mas}/\text{M}_\odot$;

M is expressed in $M_\odot$ units and π_{rel} in mas. Numerically,

$$\theta_E \simeq 0.638 \left(\frac{M}{0.5\,M_\odot}\right)^{1/2} \left(\frac{\pi_{rel}}{0.1\,mas}\right)^{1/2} mas. \qquad (6)$$

With typical values of θ_E of order of a fraction of a mas, it is impossible with classical telescopes to resolve the individual images of the source, since a fraction of an arcsecond is the usual resolution limit. But the images of the source produced by the microlens are distorted and magnified (images of the source have larger total area than the original, not lensed one), so that the total flux appear amplified during a microlensing event. In the simple case of an isolated microlens, the total magnification is given by

$$\mu = \frac{u^2 + 2}{u\sqrt{u^2 + 4}}, \qquad (7)$$

where u is the projected separation between the source and the lens in θ_E units. Since the source and the lens move with time, μ is a function of time. For a lens source rectilinear motion, such single-lens temporal magnification curves have a typical bell-shape aspect, with maximum magnification at peak that can reach many hundreds if the alignment is particularly good. The typical light curve of a planetary event is shown in Fig. 7.

After the microlensing pioneer times of decade 1993-2003, spent in improving the observing strategy and the instruments, the first microlensing exoplanet, a 2.6 M_{Jup} planet, was detected in event OGLE-2003-BLG-235/MOA-2003-BLG-53Lb. A milestone discovery was OGLE 2005-BLG-390 [4], the very first cool super-Earth with a mass of 5.5 $M_\oplus$ and semi-major axis of 2.6 AU, and discovered in 2005. Massive Jovian planets around low-mass red dwarfs have been discovered by microlensing, as OGLE-2005-BLG-071 and MOA-2009-BLG-387. The existence of such planets challenges the core accretion theory of planet formation, since it predicts that massive, Jovian planets should be rare around M dwarfs. Multi-planet systems have been discovered too, OGLE-2006-BLG-109 and OGLE-2012-BLG-0026. The first system is actually a scaled-down model of our solar system, with two planets analogs as Jupiter and Saturn, but of lower masses. Microlensing has discovered a population of free-floating planets [27], which may be as frequent as stars in the Milky Way. These events are of very shot timescale, less than two days.

With 700 alerts per year in 2009 to about 2500 in 2011, the OGLE collaboration has already quadrupled its number of alerted microlensing events. New generations of robotic telescopes equipped with wide-field cameras (KMNET, LCOGT) are joining the microlensing networks, complementing earlier-generations telescopes (OGLE IV, MOA II, PLANET, μFUN, RoboNET, MiNDSTEp) together with an increasing number of amateur telescopes. The growing number of light curves requires the implementation of high performance and automated modeling tools working in real time.

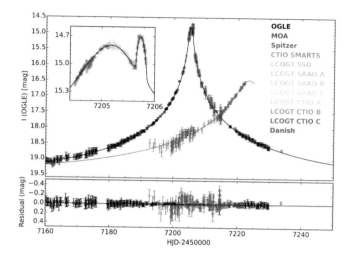

FIGURE 7. Light curve of OGLE-2015-BLG-0966, with combined data from 10 ground-based telescopes (data fitting the black solid line model). The violet curve is a light curve obtained with the Spitzer satellite. The shift between the two light curve provide important constraints on the lens-source parallax and so on the microlens physical parameters (figure from [26]).

Microlensing observations from space have been carried out with the Rosetta spacecraft to confirm the planetary nature of microlensing candidates in 2004. Since 2014, the Spitzer satellite observes microlensing events to measure the lens-source parallax, with spectacular results (see Fig. 7). Specific space-based microlensing programs have been proposed onboard the Euclid satellite (ESA, launch in 2020, additional science) and WFIRST (NASA, planning phase). Simulations show that for such space-based microlensing missions a great number of planets, including Earth-like planets, should be detected in a few months, and unprecedented constraints on the planetary mass function down to the mass of the Earth should be obtained.

2.5. Imaging

Probably the most natural observation technique in astronomy is to make an image of the object one wants to study. In exoplanet research, this technique is called direct imaging (in contrast with previous methods which relied on indirect effects). Imaging an exoplanet is not an easy task, though: exoplanets like those we know in our Solar system are billions of times fainter than their host stars and situated at extremely close angular separations from their host star. Current imaging techniques can detect an exoplanet 10^5 times fainter than its host star, at about 1 arsecond separation. Although this is already a technical prowess, it is still far from what is required to detect a planet like Jupiter around another Sun.

The direct imaging challenge is to separate the light from the planet to the light from the star, which is diffracted by the telescope. The instruments themselves are designed to block or annulate the light from the star, while the post-observation softwares are designed to distinguish between diffraction features due to the planet or to the star. Coronagraphs are the central piece of direct imaging instruments. The basic principle of imaging has been invented by Lyot in 1939: the light from the star is blocked on the optical axis by a focal plane mask (called Lyot stop), while in the pupil-plane located after, another mask blocks the light diffracted off-axis, in order to remove the starlight. Since the exoplanet is not located on the optical axis but makes a small angle with it, the light from it is not blocked by the Lyot stop and will consequently appear as a classical image. Apodizers are designed to modify the transmission of the telescope so that power of the sidelobes (off-axis) of the star light is minimum, to increase the contrast between this residual light from that of the planet (Fig. 8).

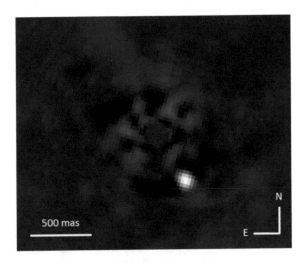

FIGURE 8. Imaging of the planetary system β Pic using the AGPM coronagraph on VLT/NACO (image in linear flux scale). The giant planet visible on the image is at 8-9 AU from the star. The technique allows to rule out the presence of additional giant companions down to orbits of 2 AU only (figure from [1]).

In practice, many optical designs have been proposed and implemented to specifically search for exoplanets, usually employing cutting edge technology. Direct imaging also requires adaptive-optic mirrors to correct for atmospheric turbulence beforehand. In fact, atmospheric turbulence is responsible for polluting the image with a halo of speckles that rapidly evolve, and mask planetary signals.

Today, the best coronagraphs can remove diffraction down to the level of 10^{-10} at separations larger than $2 - 4\lambda/D$.

Direct imaging is primarily sensitive to massive planets at wide orbits from their parent stars (greater than about 5 AU). The main (and still only) targets of this method are young exoplanets, which emit infrared light from the heat they accumulated during their formation phase and subsequently release it for a few tens million years (to be compared with, e.g., the lifetime of the Sun, about 10 billion years). The effective temperature of these young planets can reach up to 2000 K (for Jupiter today, this temperature is about 150 K). One of the main difficulty is to derive the mass from the planet's luminosity, since the latter is the only observable. Evolutionary models of the temperature as a function of age have to be used, which introduces a degree of uncertainty in the determination of the planetary masses.

In 2004, the first major discovery was that of a giant planet of 5 times the mass of Jupiter in orbit around a brown dwarf aged of about 8 million years [10]. Several detections followed, amongst them the detection of young, multiple planetary systems. In cases when the planet is sufficiently separated from its host star, direct spectroscopy observations can be performed. Planets observed this way have low gravity (as expected from the fact they still hot and slowly cooling down), and distinct atmospheric structures from brown dwarfs. In very favorable cases, individual spectral lines such as CO lines can be directly observed. These are important pieces of information when trying to understand the formation mechanisms of super-giant planets vs. brown dwarfs. When possible, direct imaging in the optical, although much more challenging since the planet-star contrast is much higher (the maximum black body emission of the planet is in the infrared, and lies in the Rayleigh-Jeans wing of the star's black body), provide important information on the albedo of the planet, and on its cloud cover.

The current approach is to develop new direct imaging instruments to specifically search for exoplanets. An adaptive optics system using 2000 actuators has been mounted on the Subaru telescope (Hawaii) to serve as a testbed for future technologies. Two instruments have been recently commissioned and have obtained first very detailed and highly contrasted images: the Gemini Planet Finder, and the SPHERE instrument at the Very Large Telescope (ESO), both located in Chile. The upcoming European Extremely Large Telescope (E-ELT), the future largest optical and near-infrared 39 m telescope is scheduled to take its first image in 2024, and should be able to reach planet-star separations of only 0.2 arcsecond. Imaging space missions are not planned, but instruments are developed onboard JWST (future space telescope in replacement of Hubble, launch planned in 2018) and prototype chronographs are studied for the future WFIRST mission (NASA).

2.6. Astrometry

Astrometry consists in measuring the position and motion of objects in the plane of the sky. It is probably the oldest branch of astronomy. Historically, many retracted discovery were announced starting at the end of the 1930s, when photographic plates became of good enough quality. One famous case is Barnard's star, for which two planetary-mass bodies of 0.7 and 0.5 Jupiter masses with periods of 12 and 20 years were proposed, and discussed many times for about 30 years. However, the sensitivity was much too low for a planet to be detected at all at that time, and even today this method has not produced any confirmed exoplanet. Nevertheless, the ESA satellite mission Gaia should report many detections in a very near future.

An astrometric orbit corresponds to the barycentric motion of a star caused by an invisible companion. This motion follows Kepler's laws, and astrometric measurements determine the value of $m^3/(M_* + m)^2$, where M_* is the host star mass and m the companion mass. Astrometry is applicable to planet searches around nearby stars of various masses and ages, with benefits for the study of the planet mass function of long-period planets, and of planets around active stars.

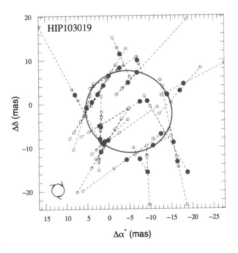

FIGURE 9. Determination of the astrometric orbit of the companion of star first detected via Doppler spectroscopy. The solid red line shows the orbital solution and open circles mark the individual Hipparcos astrometric measurements (figure from [25]).

At the time being, astrometry is used in combination with Doppler spectroscopy to refine the planet parameters (Fig. 9). In fact, only five out of seven orbital parameters are constrained by radial velocity measurements alone, and the two remaining parameters, the inclination i and the longitude of the ascending node Ω can be determined by measuring the astrometric orbit. The ESA satellite Hipparcos yielded mass upper limits on the mass of several planets, thus putting strong constraints on the inclinations of the planetary orbits. In some rare cases, it revealed that brown dwarfs or stellar companions had been mistaken for planets. The Hubble space telescope was also used for this purpose.

2.7. Other detection channels

There are many other ways to detect exoplanets, like eclipsing binary timing, radio astrometry, transit time variation, X rays emission, destabilization of debris disks... but we cannot expect these methods to contribute to the bulk of exoplanet detections. The detection of radio emission due to the interaction of the star and planet magnetospheres is an interesting possibility using imaging or astrometry. Since the first exoplanet discoveries, radio detections have been attempted numerous times, with various telescopes and at different frequencies. Although there are significant uncertainties in predicting which exoplanets are most likely to be the strongest radio emitters, until now searches have focused on the short-period planets.

3. A wealth of possible worlds

Exoplanet search campaigns have discovered a great variety of possible planets in terms of mass, orbital parameters such as eccentricity and inclination, mean densities, composition, size, or atmospheric structure to quote a few aspects. Planetary systems are shaped by formation mechanisms, different types of orbital migration, multiple body gravitational interaction, collisions, ejection of small bodies or ejection of giant planets, resonance between orbital parameters in multiple planetary systems, and many other subtle effects.

For example, the growth of planetary embryos can follow an oligarchic growth, which assumes that embryos grow from a swarm of kilometer-size planetesimals, while a more recent scenario states that the accretion of pebbles can grow an Earth-mass planet directly from cm-sized bodies in favorable conditions. These scenarios do not build up the same kinds of planets. Super-Earth planets (from 1 to 10 $M_\oplus$) can form either *in situ* by accretion of locally available material in massive disks, but they can also form at larger orbital distances and then migrate inward by interactions with the gaseous disk. The planets formed following the first recipe might not be able to capture an atmosphere, and they would all be rocky, while in the second case for the same mass, they could sustain a massive atmosphere.

The orbital migration of planets is usually an inward migration, but it some conditions in can be outward. The comparison of the number of planets located beyond or within the snow line provides important information on the efficiency of migration mechanisms. Furthermore, there are other types of migration that do not need to be gas-driven. Planet-planet scattering is another option, and migration can occur in this case after the dispersal of the gas a few million years after the birth of the star.

Many exoplanets have large eccentricities, which was unexpected. The main stages of planet formation occur in the gaseous disk, and the presence of gas very efficiently damps eccentricities and orbital inclinations, through viscous drag between the solid bodies (planetesimals or planet embryos) and the gas. The post-gaseous dynamical evolution of exo-planetary systems appears to be much more complex that was suggested by the coplanar and almost circular orbits of the planets around the Sun.

Giant planets can grow from solid planetary cores formed by accretion, in other words can use a newly-formed terrestrial planet as a core to later accrete a significant amount of gas. In this case, the planet is expected to include a fraction of solids (dust) substantially higher than that of the proto-planetary disk. Another scenario proposes the direct formation of a giant planet by gravitational instability in the proto-planetary disk, provided that the disk is massive enough. Such planets would form at large orbital distances, where the temperature is cold enough to ease gravitational collapse.

At greater masses, super-giant planets have been found to overlap with the range of masses historically attributed to brown dwarfs $(13 - 74 M_J)$. The latter are stars that could not ignite hydrogen in their core to become real stars, but still could burn deuterium for a few million years. It is not clear which have been formed by which mechanisms. Although the detection of these objects is far more easier than the detection of terrestrial planets, there are still very few discoveries because objects in this range of masses are intrinsically rare around stars – it is called the brown dwarf desert. More generally, the frequency of companions drops with increasing mass, and the frequency of super-giant planets and brown dwarfs companions to stars is estimated to be as low as 1%.

Some hot Jupiters have been found to be exaggeratedly inflated: although of Jupiter mass, some can have sizes of several times Jupiter's radius. The origin of these large radii is still unclear, but the basic reason is the atmosphere do not release its energy and increases its volume. It may be that the molecules of the upper atmosphere are ionized by the strong irradiation due to the proximity of the star, which produces electric conduction and results in heating trough ohmic resistivity. Another hypothesis is the presence of clouds, which increase the opacity of the atmosphere.

Early discoveries have shown that exoplanets can be found in orbit around one of the components of a wide-separation binary star. While the planets are always located relatively far from destabilizing resonances, the stability of the proto-planetary disks can be questioned in some cases. More recently in 2011, planets orbiting close binary systems have been discovered [11]. In this case, the exoplanet orbits the pair of stars, and is called circum-binary exoplanet.

For a subset of objects, it has been possible to measure a Rossiter-McLaughlin effect, which results from the fact that a transiting planet blocks sequentially Doppler-shifted area of the star due to its rotation. The angle between the spin of the star and the orbit of the planet can then be measured trough analysis of the radial velocity curve. In a normal case, this angle is nul, as it is the case in the Solar system. But several exoplanets do exhibit a spin-orbit misalignment, which probably results from early dynamical interactions with other planets in the formation phase.

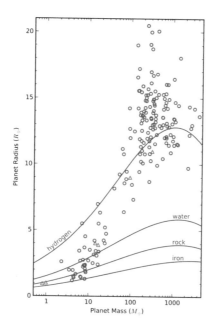

FIGURE 10. Masses and radii of exoplanets (red dots). The blue
lines are mass-radius relationships for planets with pure hydrogen,
water, rock or iron (figure from [13]).

Another recent discovery is that in the super-Earth mass regime, there are
not only heavier version of rocky planets like the Earth, but also planets with
massive atmospheric enveloppes of hydrogen and helium (Fig. 10). Intermediate
radius could indicate ocean planets with atmospheres saturated of water vapor.

Exoplanets not related to any star are called free-floating planets. They have
been discovered via direct imaging and microlensing. While for the largest objects
it may be that they formed directly by gravitational instability, for Jupiter-mass
objects it is very likely that they were ejected from their planetary system after
the gas was dispersed from the proto-planetary disk. If the system had formed two
giant planets located too close to each other, gravitational interaction may result
in the ejection of one of the planet, with the second planet moved to an eccentric
orbit. Indeed, many extrasolar planets have rather large eccentricities, although
there are many possible origins for that.

Precise spectroscopy of transiting planets have revealed the presence of
molecules in the atmosphere of giant planets, in particular water vapor [28]. In
other hot giant planets, spectroscopy resulted in flat spectra which could result
from a thick cloud cover or haze in the planetary atmosphere.

Phases of hot Jupiter planets can be sometimes observed (Fig. 5) and the wind speed of equatorial jets measured at the surface of hot Jupiters via the displacement of the hottest point on the planet's surface.

Finally, families of comets have been found around stars harboring planets [19], as it is the case in our Solar system.

4. Rule rather than exception: more exoplanets than stars in the Milky Way

Statistical studies aim at understanding the planetary populations and at constraining their frequency, beyond the observational biases which confuse the picture. For example, naively reading Fig. 2 would suggest that hot Jupiters are very numerous (many planets at large masses and very short orbit), but in reality they account for less than a few percents of the full population. Super-Earths on contrary are not very often detected, but they actually form the largest population of exoplanets known today — keeping in mind we don't know yet how frequent are Earth-like planets. We describe below a statistical study based on microlensing data.

Giant planets located at a few AUs, like Jupiter in the Solar System, were the prime targets of the first microlensing campaigns starting in 1995. The large caustic structures implied a planet detection efficiency as high as 100% for well-covered events observed at high magnification. Because of this very high detection efficiency, many giant planets should have been detected quickly, but no planets were found yet after the first five years of observations. While part of the negative result has roots the chosen observing strategy, after a few years it became clear that giant planets at large orbital distances were intrinsically rare. The hunt for extrasolar planets through microlensing revealed itself to be more challenging than initially thought. Early statistical estimations using microlensing 1995-99 data [15] led to first significant upper limits on the abundance of giant planets around red dwarfs: less than 1/3 of the lens stars had Jupiter-mass companions, while less than 2/3 of the lenses had Saturn-mass companions in the orbital range $1.5 - 4$ AU.

After 2005, the μFUN microlensing collaboration took advantage of a growing community of amateur astronomers observing microlensing events to set up a very reactive observing strategy dedicated to detect and characterize high-magnification events. During the 4 seasons 2005-08, high-magnification events were monitored as intensively as possible, independently of any evidence of a light curve anomaly. This strategy turned out to be very efficient: half of the events monitored were planetary events. In order to estimate the planet frequency from these high-magnification events, an unbiased sample of 13 high-magnification events with peak magnification greater than 200 was selected. A point on the planetary mass function could be estimated [16], $f = 0.36 \pm 0.15$ dex^{-2} per $\log q \times \log d$. This result was consistent with Doppler estimates when considering that Doppler hosts are G dwarfs rather than M dwarfs, and when planetary systems are scaled to the location of their snow line.

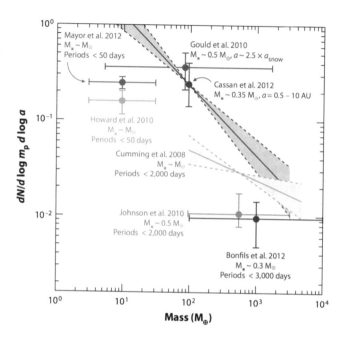

FIGURE 11. Various planetary mass function constraints from different analysis using microlensing and radial velocity data (figure from [14]).

With a database rich with about 15 years of data obtained with a worldwide network of telescopes, in 2010 the PLANET collaboration had a much higher sensitivity to low-mass planets than previous studies. The analysis was based on 1995-2010 PLANET data alerted by OGLE, using detections and non-detections [8]. In order to be combined in a meaningful statistical analysis, two critical conditions were required. First, the observing strategy should be well understood and keep homogeneous for the whole sample of events, which required that the event selection and sampling rate was chosen regardless of whether the lens harbored a planet or not. Secondly, the detections should result from exactly the same strategy than for the non-detections. It resulted from a detailed study of the statistical properties of the microlensing events that only a sub-sample of six years of data, 2002-07, satisfied these requirements. In fact, when starting its new operations in 2002, OGLE III dramatically increased its number of alerts compared to OGLE II (389 in 2002 vs. 78 alerts in 2000), which had a strong impact on the PLANET strategy. After 2007, a very open collaboration between the different microlensing teams again resulted in a dramatic change in the observing strategy. During

2002-07, PLANET monitored around 10-16% of all OGLE alerts. A power-law planetary mass function was derived from this analysis,

$$f = 10^{-0.62\pm0.22} \left(\frac{M}{95\,M_E} \right)^{-0.73\pm0.17}, \qquad (8)$$

centered on Saturn's mass. The microlensing result implied that $17^{+6}_{-9}\%$ (1/6) of stars host Jupiter-mass planets (0.3-10 M_J), while cool Neptunes (10-30 M_E) and super-Earths (5-10 M_E) are even more common, with respective abundances per star of $52^{+22}_{-29}\%$ (1/2) and $62^{+35}_{-37}\%$ (2/3). Planets around Milky Way stars are the rule rather than the exception.

Different planetary mass function estimations from microlensing and other detection techniques are shown in Fig. 11. It is important to note that they are not directly comparable, because the methods don't have the same sensitivity, and because the planets may not orbit the same types of stars.

5. Epilogue

In about two decades, we progressed from the observation of one to several hundreds of planetary systems. Many complementary techniques have led to the discovery of an astonishing diversity of foreign worlds. There are one to two hundred billion stars in the Milky Way, and even more exoplanets... how many options does it represent for the apparition of extraterrestrial life? The field of exoplanet research has open the door, but nobody knows yet what stands behind.

References

[1] Absil, O., Milli, J., Mawet, D., Lagrange, A.-M., Girard, J., Chauvin, G., Boccaletti, A., Delacroix, C., and Surdej, J.: Searching for companions down to 2 AU from β Pictoris using the L'-band AGPM coronagraph on VLT/NACO. Astron. Astrophys. **559**, L12 (2013).

[2] Backer, D. C., Foster, R. S., and Sallmen, S.: A second companion of the millisecond pulsar 1620 - 26. Nature **365**, 817–819 (1993).

[3] Bailes, M., Lyne, A. G., and Shemar, S. L.: A planet orbiting the neutron star PSR1829 - 10. Nature **352**, 311–313 (1991).

[4] Beaulieu, J.-P., Bennett, D. P., Fouqué, P., Williams, A., Dominik, M., Jørgensen, U. G., Kubas, D., Cassan, A., Coutures, C., Greenhill, J., Hill, K., Menzies, J., Sackett, P. D., Albrow, M., Brillant, S., Caldwell, J. A. R., Calitz, J. J., Cook, K. H., Corrales, E., Desort, M., Dieters, S., Dominis, D., Donatowicz, J., Hoffman, M., Kane, S., Marquette, J.-B., Martin, R., Meintjes, P., Pollard, K., Sahu, K., Vinter, C., Wambsganss, J., Woller, K., Horne, K., Steele, I., Bramich, D. M., Burgdorf, M., Snodgrass, C., Bode, M., Udalski, A., Szymański, M. K., Kubiak, M., Więckowski, T., Pietrzyński, G., Soszyński, I., Szewczyk, O., Wyrzykowski, L., Paczyński, B., Abe, F., Bond, I. A., Britton, T. R., Gilmore, A. C., Hearnshaw, J. B., Itow, Y., Kamiya, K., Kilmartin, P. M., Korpela, A. V., Masuda, K., Matsubara, Y., Motomura, M.,

Muraki, Y., Nakamura, S., Okada, C., Ohnishi, K., Rattenbury, N. J., Sako, T., Sato, S., Sasaki, M., Sekiguchi, T., Sullivan, D. J., Tristram, P. J., Yock, P. C. M., and Yoshioka, T.: Discovery of a cool planet of 5.5 Earth masses through gravitational microlensing. Nature **439**, 437–440 (2006).

[5] Butler, R. P., and Marcy, G. W.: A Planet Orbiting 47 Ursae Majoris. Astrophys. J. **464**, L153 (1996).

[6] Campbell, B., Walker, G. A. H., and Yang, S.: A search for substellar companions to solar-type stars. Astrophys. J. **331**, 902–921 (1988).

[7] Cassan, A.: Extrasolar planets detections and statistics through gravitational microlensing. Master's thesis, Sorbonne Universités, UPMC Univ Paris 06, CNRS UMR 7095, Institut d'Astrophysique de Paris, F-75014, Paris, France, cassan@iap.fr, October 2014.

[8] Cassan, A., Kubas, D., Beaulieu, J.-P., Dominik, M., Horne, K., Greenhill, J., Wambsganss, J., Menzies, J., Williams, A., Jørgensen, U. G., Udalski, A., Bennett, D. P., Albrow, M. D., Batista, V., Brillant, S., Caldwell, J. A. R., Cole, A., Coutures, C., Cook, K. H., Dieters, S., Prester, D. D., Donatowicz, J., Fouqué, P., Hill, K., Kains, N., Kane, S., Marquette, J.-B., Martin, R., Pollard, K. R., Sahu, K. C., Vinter, C., Warren, D., Watson, B., Zub, M., Sumi, T., Szymański, M. K., Kubiak, M., Poleski, R., Soszynski, I., Ulaczyk, K., Pietrzyński, G., and Wyrzykowski, Ł.: One or more bound planets per Milky Way star from microlensing observations. Nature **481**, 167–169 (2012).

[9] Charbonneau, D., Brown, T. M., Latham, D. W., and Mayor, M.: Detection of Planetary Transits Across a Sun-like Star. Astrophys. J. **529**, L45–L48 (2000).

[10] Chauvin, G., Lagrange, A.-M., Dumas, C., Zuckerman, B., Mouillet, D., Song, I., Beuzit, J.-L., and Lowrance, P.: A giant planet candidate near a young brown dwarf. Direct VLT/NACO observations using IR wavefront sensing. Astron. Astrophys. **425**, L29–L32 (2004).

[11] Doyle, L. R., Carter, J. A., Fabrycky, D. C., Slawson, R. W., Howell, S. B., Winn, J. N., Orosz, J. A., Prcaronsa, A., Welsh, W. F., Quinn, S. N., Latham, D., Torres, G., Buchhave, L. A., Marcy, G. W., Fortney, J. J., Shporer, A., Ford, E. B., Lissauer, J. J., Ragozzine, D., Rucker, M., Batalha, N., Jenkins, J. M., Borucki, W. J., Koch, D., Middour, C. K., Hall, J. R., McCauliff, S., Fanelli, M. N., Quintana, E. V., Holman, M. J., Caldwell, D. A., Still, M., Stefanik, R. P., Brown, W. R., Esquerdo, G. A., Tang, S., Furesz, G., Geary, J. C., Berlind, P., Calkins, M. L., Short, D. R., Steffen, J. H., Sasselov, D., Dunham, E. W., Cochran, W. D., Boss, A., Haas, M. R., Buzasi, D., and Fischer, D.: Kepler-16: A Transiting Circumbinary Planet. Science **333**, 1602 (2011).

[12] Duquennoy, A., and Mayor, M.: Multiplicity among solar-type stars in the solar neighbourhood. II - Distribution of the orbital elements in an unbiased sample. Astron. Astrophys. **248**, 485–524 (1991).

[13] Fischer, D. A., Howard, A. W., Laughlin, G. P., Macintosh, B., Mahadevan, S., Sahlmann, J., and Yee, J. C.: Exoplanet Detection Techniques. Protostars and Planets VI, pages 715–737 (2014).

[14] Gaudi, B. S.: Microlensing Surveys for Exoplanets. ARA&A **50**, 411–453 (2012).

[15] Gaudi, B. S., et al: Microlensing constraints on the frequency of jupiter-mass companions: Analysis of 5 years of planet photometry. Astrophys. J. **566**, 463–499 (2002).

[16] Gould, A., Dong, S., Gaudi, B. S., Udalski, A., Bond, I. A., Greenhill, J., Street, R. A., Dominik, M., Sumi, T., Szymański, M. K., Han, C., Allen, W., Bolt, G., Bos, M., Christie, G. W., DePoy, D. L., Drummond, J., Eastman, J. D., Gal-Yam, A., Higgins, D., Janczak, J., Kaspi, S., Kozłowski, S., Lee, C.-U., Mallia, F., Maury, A., Maoz, D., McCormick, J., Monard, L. A. G., Moorhouse, D., Morgan, N., Natusch, T., Ofek, E. O., Park, B.-G., Pogge, R. W., Polishook, D., Santallo, R., Shporer, A., Spector, O., Thornley, G., Yee, J. C., μFUN Collaboration, Kubiak, M., Pietrzyński, G., Soszyński, I., Szewczyk, O., Wyrzykowski, Ł., Ulaczyk, K., Poleski, R., OGLE Collaboration, Abe, F., Bennett, D. P., Botzler, C. S., Douchin, D., Freeman, M., Fukui, A., Furusawa, K., Hearnshaw, J. B., Hosaka, S., Itow, Y., Kamiya, K., Kilmartin, P. M., Korpela, A., Lin, W., Ling, C. H., Makita, S., Masuda, K., Matsubara, Y., Miyake, N., Muraki, Y., Nagaya, M., Nishimoto, K., Ohnishi, K., Okumura, T., Perrott, Y. C., Philpott, L., Rattenbury, N., Saito, T., Sako, T., Sullivan, D. J., Sweatman, W. L., Tristram, P. J., von Seggern, E., Yock, P. C. M., MOA Collaboration, M. Albrow, Batista, V., Beaulieu, J. P., Brillant, S., Caldwell, J., Calitz, J. J., Cassan, A., Cole, A., Cook, K., Coutures, C., Dieters, S., Dominis Prester, D., Donatowicz, J., Fouqué, P., Hill, K., Hoffman, M., Jablonski, F., Kane, S. R., Kains, N., Kubas, D., Marquette, J.-B., Martin, R., Martioli, E., Meintjes, P., Menzies, J., Pedretti, E., Pollard, K., Sahu, K. C., Vinter, C., Wambsganss, J., Watson, R., Williams, A., Zub, M., PLANET Collaboration, Allan, A., Bode, M. F., Bramich, D. M., Burgdorf, M. J., Clay, N., Fraser, S., Hawkins, E., Horne, K., Kerins, E., Lister, T. A., Mottram, C., Saunders, E. S., Snodgrass, C., Steele, I. A., Tsapras, Y., RoboNet Collaboration, Jørgensen, U. G., Anguita, T., Bozza, V., Calchi Novati, S., Harpsøe, K., Hinse, T. C., Hundertmark, M., Kjærgaard, P., Liebig, C., Mancini, L., Masi, G., Mathiasen, M., Rahvar, S., Ricci, D., Scarpetta, G., Southworth, J., Surdej, J., Thöne, C. C., and MiNDSTEp Consortium: Frequency of Solar-like Systems and of Ice and Gas Giants Beyond the Snow Line from High-magnification Microlensing Events in 2005-2008. Astrophys. J. **720**, 1073–1089 (2010).

[17] Henry, G. W., Marcy, G. W., Butler, R. P., and Vogt, S. S.: A Transiting "51 Peg-like" Planet. Astrophys. J. **529**, L41–L44 (2000).

[18] Hewish, A.: Pulsars. Scientific American **219**, 25–35 (1968).

[19] Kiefer, F., Lecavelier des Etangs, A., Boissier, J., Vidal-Madjar, A., Beust, H., Lagrange, A.-M., Hébrard, G., and Ferlet, R.: Two families of exocomets in the β Pictoris system. Nature **514**, 462–464 (2014).

[20] Latham, D. W., Stefanik, R. P., Mazeh, T., Mayor, M., and Burki, G.: The unseen companion of HD114762 - A probable brown dwarf. Nature **339**, 38–40 (1989).

[21] Marcy, G. W., and Benitz, K. J.: A search for substellar companions to low-mass stars. Astrophys. J. **344**, 441–453 (1989).

[22] Marcy, G. W., and Butler, R. P.: A Planetary Companion to 70 Virginis. Astrophys. J. **464**, L147 (1996).

[23] Mayor, M., and Queloz, D.: A Jupiter-Mass Companion to a Solar-Type Star. Nature **378**, 355 (1995).

[24] Paczynski, B.: Gravitational microlensing by the galactic halo. Astrophys. J. **304**, 1–5 (1986).

[25] Sahlmann, J., Ségransan, D., Queloz, D., Udry, S., Santos, N. C., Mayor, M., Naef, D., Pepe, F., and Zucker, S.: Search for brown-dwarf companions of stars. Astron. Astrophys. **525**, A95 (2011).

[26] Street, R. A., Udalski, A., Calchi Novati, S., Hundertmark, M. P. G., Zhu, A., Yee, J., Tsapras, Y., Bennett, D. P., RoboNet Project, T., Consortium, Gould, Jorgensen, U. G., Dominik, M., Andersen, M. I., Bachelet, E., Bozza, V., M., D. M., Burgdorf, M. J., Cassan, A., Ciceri, S., D'Ago, G., Dong, S., Evans, D., ich, S.-H., Harkonnen, H., Hinse, T. C., Horne, K., Figuera Jaimes, R., Kains, N., Ku, E., Korhonen, H., Kuffmeier, M., Mancini, L., Menzies, J., Mao, S., Peixinho, s, Popovas, A., Rabus, M., Rahvar, S., Ranc, C., Tronsgaard Rasmussen, R., Scarpetta, G., Schmidt, R., Skottfelt, J., Snodgrass, C., Southworth, J., Steele, I. A., Surdej, J., Unda-Sanzana, E., Verma, P., von Essen, C., Wambsganss, J., Wang, Y.-B., Wertz, O., OGLE Project, T., Poleski, R., Pawlak, M., Szymanski, M. K., Skowron, J., Mroz, P., Kozlowski, S., Wyrzykowski, L., Pietrukowicz, P., Pietrzynski, G., Soszynski, I., Ulaczyk, K., The Spitzer Team Beichman, C., Bryden, G., Carey, S., Gaudi, B. S., Henderson, C., Pogge, R. W., Shvartzvald, Y., The MOA Collaboration, Abe, F., Asakura, Y., Bhattacharya, A., Bond, I. A., Donachie, M., Freeman, M., Fukui, A., Hirao, Y., Inayama, K., Itow, Y., Koshimoto, N., Li, M. C. A., Ling, C. H., Masuda, K., Matsubara, Y., Muraki, Y., Nagakane, M., Nishioka, T., Ohnishi, K., Oyokawa, H., Rattenbury, N., Saito, T., Sharan, A., Sullivan, D. J., Sumi, T., Suzuki, D., Tristram, P. J., Wakiyama, Y., Yonehara, A., KMTNet Modeling Team Han, C., Choi, J.-Y., Park, H., Jung, Y. K., and Shin, I.-G.: Spitzer Parallax of OGLE-2015-BLG-0966: A Cold Neptune in the Galactic Disk. ArXiv e-prints, (2015).

[27] Sumi, T., Kamiya, K., Bennett, D. P., Bond, I. A., Abe, F., Botzler, C. S., Fukui, A., Furusawa, K., Hearnshaw, J. B., Itow, Y., Kilmartin, P. M., Korpela, A., Lin, W., Ling, C. H., Masuda, K., Matsubara, Y., Miyake, N., Motomura, M., Muraki, Y., Nagaya, M., Nakamura, S., Ohnishi, K., Okumura, T., Perrott, Y. C., Rattenbury, N., Saito, T., Sako, T., Sullivan, D. J., Sweatman, W. L., Tristram, P. J., Udalski, A., Szymański, M. K., Kubiak, M., Pietrzyński, G., Poleski, R., Soszyński, I., Wyrzykowski, L., Ulaczyk, K., and Microlensing Observations in Astrophysics (MOA) Collaboration: Unbound or distant planetary mass population detected by gravitational microlensing. Nature **473**, 349–352 (2011).

[28] Tinetti, G., Vidal-Madjar, A., Liang, M.-C., Beaulieu, J.-P., Yung, Y., Carey, S., Barber, R. J., Tennyson, J., Ribas, I., Allard, N., Ballester, G. E., Sing, D. K., and Selsis, F.: Water vapour in the atmosphere of a transiting extrasolar planet. Nature **448**, 169–171 (2007).

[29] Winn, J. N.: *Exoplanet Transits and Occultations.* Exoplanets, Seager, S., editor, 55-77, dec. 2010, http://cdsads.u-strasbg.fr/abs/2010exop.book...55W, Provided by the SAO/NASA Astrophysics Data System.

[30] Wolszczan, A.: Discovery of pulsar planets. New Astron. **56**, 2–8 (2012).

Arnaud Cassan
Université Pierre et Marie Curie, Institut d'Astrophysique de Paris
98bis boulevard Arago, 75014 Paris, France
e-mail: cassan@iap.fr

Printed in the United States
by Baker & Taylor Publisher Services